T0213989

Undergraduate Topics in Computer Science

'Undergraduate Topics in Computer Science' (UTiCS) delivers high-quality instructional content for undergraduates studying in all areas of computing and information science. From core foundational and theoretical material to final-year topics and applications, UTiCS books take a fresh, concise, and modern approach and are ideal for self-study or for a one- or two-semester course. The texts are all authored by established experts in their fields, reviewed by an international advisory board, and contain numerous examples and problems, many of which include fully worked solutions.

The UTiCS concept relies on high-quality, concise books in softback format, and generally a maximum of 275–300 pages. For undergraduate textbooks that are likely to be longer, more expository, Springer continues to offer the highly regarded Texts in Computer Science series, to which we refer potential authors.

More information about this series at https://link.springer.com/bookseries/7592

K. Erciyes

Algebraic Graph Algorithms

A Practical Guide Using Python

Springer

K. Erciyes
Software Engineering Department
Maltepe University
Maltepe, Istanbul, Turkey

ISSN 1863-7310   ISSN 2197-1781   (electronic)
Undergraduate Topics in Computer Science
ISBN 978-3-030-87885-6   ISBN 978-3-030-87886-3   (eBook)
https://doi.org/10.1007/978-3-030-87886-3

This Springer imprint is published by the registered company Springer Nature Switzerland AG
The registered company address is: Gewerbestrasse 11, 6330 Cham, Switzerland

To my parents for taking so much effort to raise a large family

Preface

Graphs are discrete structures that find many applications such as modeling computer networks, social networks and biological networks. Graph theory is centered around studying graphs and there has been an unprecedented growth in this branch of mathematics over the last few decades, mainly due to the realization of numerous applications of graphs in real-life.

This book is about the design and analysis of algebraic algorithms to solve graph problems. The algebraic way of analyzing graph problems can be viewed from the angles of group theory and linear algebra. We will mostly use linear algebraic methods which commonly make use of matrices associated with graphs in search of suitable graph algorithms. There are few benefits to be gained by this approach; first of all, many results from matrix algebra theory become readily available, for matrix analysis of graphs. Secondly, various matrix algorithms such as matrix multiplication are readily available, enabling easiness in coding. Moreover, methods for parallel matrix computations are well known and various libraries for this purpose are available which provides a convenient way for parallelizing the algorithms.

The algebraic nature of graphs is a vast theoretical topic with many new and frequent results. For this reason, the level of exposure required careful consideration while forming the detailed topics of the book. At one end; we have rich, resourceful and sometimes quite complicated matrix algebra theory that can be used in the analysis of graphs which does not always result in practical graph algorithms; and at the other extreme, one can frequently devise an algebraic version of a classical graph algorithm by using the matrices that represent graphs. We tried to stay somewhere in between by reviewing main algebraic results that are useful in designing practical graph algorithms and on the other hand, mostly using graph matrices to solve graph problems. The extent of exposure to parallel processing was another decision and after briefly reviewing the basic theory on parallel processing, we provide practical hints for parallel processing associated with algebraic algorithms where possible. Matrix multiplication is at the core of majority of algebraic graph algorithms and any such algorithm may be parallelized conveniently, at least partly, using parallel matrix multiplication methods we describe. Thus, obtaining parallel version of the algorithms reviewed becomes a trivial task in most cases. In summary, the focus of the book is on practical algebraic graph algorithms using results from matrix algebra rather than algebraic study of graphs.

The intended audience for this book is the senior/graduate students of computer science, electrical and electronic engineering, bioinformatics, and any researcher or a person with background in discrete mathematics, basic graph theory and algorithms. There is a Web page for the book to keep errata Python code and other material at: http://ube.ege.edu.tr/˜erciyes/AGA/erciyes/AGA/.

Python Implementation
For almost all algorithms, we provide Python programming language code that can be modified and tested for various inputs. Python is selected for its simplicity, efficiency and rich library routines and hence the name of the book. The code for an algorithm is not optimized and brevity is forsaken for clarity in many cases. In various algorithms, we use different data structures and methods for similar problems to show alternative implementations. The codes are tested for various sample graphs, however; as it frequently happens with any software, errors are possible and I would be happy to know any bugs by e-mail at kayhanerayes@maltepe.edu.tr.

I would like to thank senior/graduate students at Ege University, University of California Davis, California State University San Marcos, Izmir Institute of Science and Technology, Izmir University, Üsküdar University and Maltepe University in chronological order, who have taken courses related to graph theory and algorithms, sometimes under slightly different names, for their valuable feedback when parts of the material covered in the book was presented during lectures. I would also like to thank Springer senior editor Wayne Wheeler for his continuous help and encouragement throughout the writing of the book.

Maltepe, Turkey   K. Erciyes

Contents

Introduction

1

Abstract

This chapter serves as an informal introduction to the basic concepts used in the book, which are graphs and matrices, along with implementation ideas. It contains a very brief introduction to Python programming language, main challenges in algebraic graph algorithms and a detailed description of the contents of the book.

1.1 Graphs

A graph is a discrete structure consisting of vertices and edges. Graphs can be used to model many real-life phenomenon, for example, vertices of a graph may represent cities and edges are used to show the roads between the cities. In bioinformatics, protein networks inside a cell may be represented by a graph with vertices showing the proteins and edges the interaction between them. A graph is denoted by $G = (V, E)$ where V is the vertex set and E is the edge set of the graph. Figure 1.1 displays a graph with a vertex set $V = \{a, b, c, d, e\}$ and and edge set $E = \{(a, b), (b, c), (c, d), (d, e), (b, d), (a, d), (b, e), (a, e)\}$.

A graph may have weights associated with its edges, representing some parameter. For example, weights in such a *weighted graph* may represent the distances between cities in a road map. The shortest path between vertices a and b in the graph of Fig. 1.1 is $a - e - b$. An *algorithm* is a sequence of instructions to solve a problem. An algorithm should be correct, have favorable performance and should terminate. A *graph algorithm* is used to solve a problem related to a graph. Finding the shortest distance between two vertices of a graph may be one such problem.

Our aim in this book is to solve well-known graph problems using algebraic methods which are algorithms that involve matrices related to graphs.

© Springer Nature Switzerland AG 2021

K. Erciyes, *Algebraic Graph Algorithms*, Undergraduate Topics in Computer Science, https://doi.org/10.1007/978-3-030-87886-3_1

Fig. 1.1 An example
undirected graph

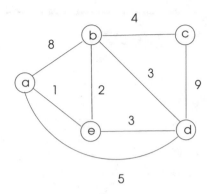

1.2 Matrices

An $m \times n$ matrix is a data structure with $m \cdot n$ real numbers arranged in m rows and
n columns. The element at ith row and jth column of a matrix A is denoted by a_{ij}.
The following matrix has 3 rows and 4 columns with $a_{2,1} = 7$ for example.

$$\begin{bmatrix} 1 & 0 & 3 & 2 \\ 5 & 2 & 1 & 4 \\ 0 & 7 & 2 & 6 \end{bmatrix}$$

Matrices find a variety of applications such as solving linear equations and are used
extensively to solve large scientific problems. Various operations such as addition,
subtraction and multiplication of matrices are defined.

A graph may be represented by a matrix called the adjacency matrix. The a_{ij} of
adjacency matrix A of a graph G is equal to 1 if there is an edge between vertices i
and j. This simple structure may be used to define a graph precisely and we will see
that many graph algorithms may be realized by some definite operations of this and
other graph related matrices.

1.3 Algebraic Graph Algorithms

We will call the type of graph algorithms that make use of a matrix related to the graph
algebraic graph algorithms which forms the basis of this book. There are few benefits
of using this approach to solve a graph problem. First of all, the matrix operations
in matrix algebra in the general sense is a mature topic with many readily available
algorithms [1]. Parallel processing is used to distribute parts of a computation or data
to a number of individual processing elements to achieve faster operation. This type
of processing is needed especially when data used in computation is large. We can
have shared memory parallel processing where the processors share data through a
global memory accessible to all of them, or distributed memory parallel processing
with each processor having local memory exchanging data using messages only.

Based on the foregoing, our procedure to solve a graph problem commonly consists of the following steps.

1. Represent the graph using a matrix. This matrix is commonly the adjacency matrix.
2. Show how to implement the required graph algorithm using this matrix as the *only* input, commonly using pseudocode.
3. Provide Python code for the algorithm with the obtained output.
4. Provide hints on how to convert the sequential algorithm into a shared memory or a distributed memory parallel algorithm.

We will be using basic matrix operations such as addition or multiplication of matrices to solve a given graph problem. The parallel algorithms for these basic operations already exist in literature [2] and describing where they can be used in the algorithm will be adequate to show how to parallelize the graph algorithm in step 4 above.

We will be implementing Python to demonstrate workable code for most of the algorithms presented in this book mainly because Python has a rich set of libraries that allow performing some low-level operations conveniently.

1.4   Python Language

Python is a general purpose programming language that finds numerous application areas from network programming, cybersecurity, artificial intelligence, scientific computing to machine learning. It has a simple syntax making it easier to learn than many other programming languages. It can work on various platforms including Windows and various Linux derivatives and can be implemented as a procedural, functional or object-oriented language.

Python was designed by Guido van Rossum in the 1980s with the aim of providing a language that does not have too much freedom but also is not very restrictive. A very strong side of Python is the existence of many libraries for diverse applications. The reason for choosing Python over various other languages to demonstrate algebraic graph algorithms in this book was basically this property; most of the low level routines were readily available from various libraries. Thus, our focus could be oriented towards higher level concepts and methods to design algorithms at a moderately higher level.

There is no need to specify a data type in Python, the interpreter understands the data type from the context. Python runs on an interpreter system where code is run as soon as it is written. The indentation and white spaces are used to define block structures. Three types of data structures in Python that we will be using frequently are *arrays*, *lists* and *dictionaries*. An array is a homogenous data structure that is commonly used to store numbers as in various other languages. A list is a data structure like an array, its elements can be accessed using indices. However, the elements of a list may be heterogenous, making it suitable to store different types of elements. A dictionary on the other hand consists of key-value pairs.

Python has many libraries targeted at different applications. Out of these libraries, we will make frequent use of the following while designing and implementing algebraic graph algorithms. We will provide a more detailed coverage of Python and its libraries in Chap. 2.

- *numpy*: Numerical Python (NumPy or numpy) is a Python library package basically for matrices and arays. It includes methods to create, manipulate and operate on matrices. It needs to be imported prior to implementation. We will be using this module frequently when performing matrix operations such as multiplication.
- *scipy*: Scientific Python (SciPy) is used for scientific computing. It has modules such as linear algebra, optimization and image processing. It is built on top of *numpy* and uses array structure from this module. We will use this module for some special operations such as finding the eigenvalues of a matrix.
- *math*: The math module is a built-in module of Python to be used for basic math functions such as sine or cosine of a number.
- *multiprocessing*: This module is used for shared memory parallel processing.
- *mpi4py*: The message passing interface (MPI) is a widely adopted standard for parallel processing. The Python library for MPI called *mpi4py* is used for distributed memory parallel processing.

1.5 Challenges

There are a number of challenges in algebraic graph design and implementations. A list of possible challenges may be listed as follows.

- First and foremost, given a graph problem X, can we reduce or convert it to an algebraic one? We will see that some problems do not naturally lend to algebraic solutions. We will frequently make use of some algebraic structures such as the adjacency matrix and incidence matrix of a graph to solve such problems.
- Even if we can solve a graph problem using matrices, is the solution optimal? We will find that in some cases, the algebraic algorithm has worse time complexity than the non-algebraic one.
- Is parallel implementation of the solution algorithm possible? If the designed algorithm contains some basic matrix operation and that operation can be performed in parallel, then we can deduce that there is at least some inherent parallelism in the algorithm. We will find that matrix-vector multiplication and matrix-matrix multiplication are very common subroutines to use in the design and implementation of algebraic graph algorithms. Thus, most of the algorithms designed will comprise some parallel code.

1.6   Outline of the Book

We have two parts in the book: Background and Basic Algorithms.

1.6.1   Background

The first four chapters after the Introduction Chapter is aimed to form the necessary background for algebraic graph algorithms. The prerequisite for these algorithms are matrices and matrix computations, parallel matrix computations and the relations of graphs, matrices and matroids. With these in mind, we have Chap. 2 as a short introduction to Python programming language. Chapter 3 is a dense review of matrix algebra in which we define matrices, properties of matrices, types of matrices and various matrix computations in terms of algorithms. For most of matrix computations, we show how to implement them using some library function of Python and occasionally in pseudocode. These algorithms and Python functions will serve as building blocks of the algebraic graph algorithms we will implement in the rest of the book.

Chapter 4 in this part starts with basic graph terminology, definitions, types, operations on graphs. Our attention here is the matrix representation of graphs as we will be using these structures in almost every algorithm to be described. The last section of this chapter is a review of matroid structure which can be used to solve some graph problems efficiently. The common properties of graphs, matrices and matroids are also described to aid design of some graph algorithms.

Chapter 5 is a review of parallel processing concepts and how to implement parallel processing using Python. We describe shared memory and distributed memory parallel processing methods with emphasis on distributed memory systems and their communication patterns. Our focus here is on parallel matrix multiplications in a distributed system as we will frequently employ these procedures. Lastly, we investigate two widely used Python parallel processing environments: *multiprocessing* for shared memory and *mpi4py* for distributed memory parallel processing using the message passing interface (MPI) standard. We show how to implement matrix multiplication in parallel using both approaches. The last section in this chapter is the sparse matrices which have many zeros as their elements; we state formats for storage of these matrices and basic multiplication of them using Python.

1.6.2   Basic Graph Algorithms

We start basic graph algorithms with algorithms to build a spanning tree of a graph, both in sequential and parallel forms in Chap. 6. The main topic in this chapter is building a minimum spanning tree (MST) of a weighted graph. We describe Jarnik-Prim algorithm, Kruskal's algorithm, and Boruvka's algorithm for this purpose with sequential implementations in Python and parallel implementations using the matrix representations of the graphs.

Finding shortest paths in an unweighted/weighted graph has numerous applications and this problem is our main investigation topic in Chap. 7. We start with the algebraic Breadth-First-Search algorithm which is quite different then the non-algebraic version. We show its implementation in Python and provide hints for parallel execution. Dijkstra's algorithm, Bellman-Ford algorithm for single-source-shortest-paths and Floyd-Warshall algorithms for all-pairs-shortest-paths are also reviewed in this chapter with similar procedure; pseudocode, Python implementation and hints for parallel processing. Lastly, Warshall's algorithm for transitive closure problem is included with implementation code in Python.

Chapter 8 is about two fundamental problems in graph theory; connectivity and matching. A graph is connected if it is possible to reach every vertex from every other vertex, that is, there is a path between each vertex pair. We review algorithms to test whether a graph is connected and the number and identifier of vertices of components which are disjoint subgraphs of a graph. A matching is non-adjacent edge set in a graph and the second part of this chapter is about algebraic matching algorithms.

Chapter 9 in this part is on detecting subgraphs with some defined property of a graph. We first look at independent sets which are sets of non-adjacent vertices in a graph. A dominating set is a subset of vertices of a graph such that any vertex of the graph is either in this set or adjacent to a vertex in the set. A vertex cover is again a subset of vertices in a graph where every edge of the graph is incident to at least one vertex in this set. The last problem we investigate is the vertex coloring problem in which a color in the form of an integer is assigned to a vertex with the requirement that the color of a vertex is different than any of its neighbors. None of these problems have solutions in polynomial time and thus, heuristics which are common sense rules are used to find suboptimal solutions. We review heuristic algorithms for these problems and show the implementations in Python.

Chapter 10 is dedicated to problems of large graphs. Many real-life networks are large with tens of thousands of vertices and hundreds of thousands of edges. These networks are called *complex networks* and they have interesting and somehow unexpected properties. The distance between any two vertices in a complex networks is small and thus, they are called *small-world* networks. Also, a few number of vertices in such networks have many connections where the rest of vertices commonly have very few connections. This property is known as *scale-free property* of a complex network. We need new methods to analyze graphs when they are used to represent a complex network. The first part of this chapter is a dense review of graph analysis methods with related Python functions to realize the analysis. In the last part of this chapter, we review methods to form and analyze Kronecker graphs. A Kronecker product of the adjacency matrix of a graph with itself a number of times provides a graph called the *Kronecker graph* with two basic complex network properties: small-world and scale-free networks. We can therefore use this property to generate complex networks superficially and then analyze their properties.

The main topic of Chap. 11 is the two related problems: graph partitioning and clustering. Graph partitioning of an unweighted aims to divide a graph into balanced partitions with similar number of vertices in each partition and a minimum total number of edges between the partitions. In a weighted graph, the goal of partitioning is to have similar total weight of vertices in each partition when vertices have weights and to have a minimum total weight of edges between partitions when edges have weights. We review a simple BFS-based algorithm and a spectral bisection algorithm that uses eigenvalues and eigenvectors of the Laplacian matrix of the graph and multilevel graph partitioning algorithms that contract a graph recursively, divide the graph into partitions and then project the partitions back to the original graph. Graph clustering is basically discovering the dense regions in a graph and has many applications such as finding close friends in a social network. We review MST-based and core-based clustering algorithms in the second part of this chapter.

References

1. G.H. Golub, G.F. Van Loan, *Matrix Computations* (Johns Hopkins Studies in the Mathematical Sciences), 4th edn. (2013)
2. A. Gupta, G. Karypis, V. Kumar, A. Grama, *Introduction to Parallel Computing, Design and Analysis of Algorithms*. Pearson College Div; Subsequent edition (January 1, 2003) (2003)

Part I
Background

A Short Review of Python

2

Abstract

We review basic Python features with emphasis on the properties we will use to implement algebraic graph algorithms in this chapter. We start with main data structures which are the lists, arrays, dictionaries and sets. We then review control flow methods with decision and loop structures. The last part of the chapter contains module descriptions, input/output and standard library features.

2.1 Introduction

Python is a general purpose language that is both object-oriented, procedural and functional. The current Python version is 3.x and our descriptions and coding refer to this version. In its simplest form, the Python interpreter evaluates inputs line by line as in the example below where an arithmetic expression is evaluated first and a variable a is assigned the value of 4 s.

```
>>> 5 * (3 + 4)
35
>>> a = 4
4
```

Python statements do not end with a special character such as semicolon, a new line simply means a new statement. A backslash (\) at the end of a line is used if a statement expands to more than one line. Indentation is used to define a block of statements. Indentation within a block should have the same depth within a block of statements but the length of depth may vary among different blocks.

Python when invoked by typing *python* from a terminal in UNIX environment responds by three consecutive greater ($>>>$) signs as prompt and the interpreter is exited by control-D. We can save a Python program as a file with *py* extension and running this program, say example.py, can be done in UNIX environment by,

© Springer Nature Switzerland AG 2021   11
K. Erciyes, *Algebraic Graph Algorithms*, Undergraduate Topics in Computer Science,
https://doi.org/10.1007/978-3-030-87886-3_2

```
python example.py
```

The first line of a Python program is typically the following statement which tells the shell where to find the Python interpreter.

```
#!/usr/local/bin/python
```

Assignment in Python is done by the "$=$" operator and comparison is implemented by double equal sign ($==$). The logical operators are the words *and*, *or*, *not*. Variable types do not need to be declared prior to usage, their types are determined dynamically during assignment. Multiple assignments are possible as in the following code.

```
>>> a, b = 3, 4
>>> a
3
>>> b
4
```

The basic way of getting input from the user is the *input* command and we can display a message with this command. The basic output command is *print* which can be used to output messages and values of variables etc. When the needed input is a numeric value, we need to convert the input string to its numeric value by the *eval* statement to perform some arithmetic operation as shown in the code below. Comments in python start with the # sign.

```
1   name = input('What is your name? ') # input name
2   year= eval(input('and your date of birth? ')) # convert input
3   print('Hello ', name)
4   print('You are', 2021-year, 'old')
5   >>>
6   What is your name? Cem
7   and your date of birth? 2001
8   Hello  Cem
9   You are 20 old
```

Python has numerous libraries which is one of the reasons that make Python very popular. For example, a random integer may be generated by importing the *random* class and then generating a number by using the method *randint* from this class as shown below.

```
>>> import random
>>> x = random.randint(1,10)
>>> print(x)
2
```

2.2 Data Structures

Python has numbers and strings as basic data types. Numbers are created when data is assigned to them, in other words, there is no need to declare type of a variable before assignment as stated. It is possible to convert a data type to another, for example $int(a)$ converts a to an integer and $float(a)$ converts a to a floating point number.

```
>>> a = 32.5
>>> print(int(a))
32
>>> b = 28
>>> print(float(b))
28.0
```

There exists various other higher level data structures used in Python, out of these structures, we will be using lists, arrays, dictionaries and sets frequently. Brief description of these structure is introduced in the following sections.

2.2.1   Strings

Strings are the basic data types for operation on texts. A string may be created by enclosing the text in single or double quotes. A string that expands to multiple lines is enclosed within three quotes. The length of a string may be obtained by the *len* method, two strings may be concatenated by the + operator and a string may be repeated by the * operator. The following example demonstrates these concepts.

```
1  s1 = 'Life '
2  s2 = 'is beatiful'
3  s3 = s1 + s2
4  s4 = s1 * 3
5  print(s3)
6  print(s4)
7  >>>
8  Life is beatiful
9  Life Life Life
```

The *lower* method converts every character of a string to lowercase and *upper* converts characters of a string to uppercase as in the code below.

```
1  s = 'Hello World'
2  print(s.lower())
3  print(s.upper())
4  >>>
5  hello world
6  HELLO WORLD
```

2.2.2   Lists

A list is a commonly used data structure that can contain elements of different types. For example, the following statement declares a list with three elements: an integer, a string and a float number.

```
1 = [3,'hello',0.34]
```

An element of a list may be accessed using its index, for example, printing *l*[2] will yield a value of 0.34. The following built-in functions provide frequent operations we will be using on lists:

- *len(list)*: Returns the number of elements in the list.
- *sum(list)*: Returns the sum of number of elements in the list when list contains numeric values.
- *min(list)*: Returns the minimum element in the list.
- *max(list)*: Returns the maximum element in the list.

Some commonly used list methods are as follows.

- *list.apend(a)*: Appends element a to the end of the list.
- *list.extend(l)*: Extends the list by the second list *l*.
- *list.reverse(l)*: Reverses the elements of the list.
- *list(s)*: Converts the sequence *s* to a list.
- *list.pop( )*: pops the last element from a list.
- *list.remove(a)*: Removes element *a* from a list.
- *list.insert(pos,a)*: Inserts element *a* in position *pos* in the list.
- *list.index(a)*: Returns the index of the first occurrence of element *a* in the list.
- *list.count(a)*: Counts the number of occurrences of element 'a' in the list.

Lists may be nested by creating another list within a list using square brackets. The inner list may be accessed as an element of the outer list using the index of its position and any element in the inner list may be accessed using its index in the inner list as the second index. The following example shows nesting lists inside a list, for example [6, 7] is nested inside the list [[6, 7], 8] which is included in the main list *l*.

```
1  1 = [1, 2, 3, [4, 5],[[6, 7], 8]]
2  print(1)
3  print(1[3][0])
4  print (1[4][0][1])
5  >>>
6  [1, 2, 3, [4, 5], [[6, 7], 8]]
7  4
8  7
```

The main operators that can be used with lists are + for addition, * for repetition as shown in the following example.

```
1  11 = [1, 2, 3]
2  12 = [4, 5, 6]
3  13 = 11 + 12
4  14 = 11 * 4
5  print("11: ",11)
```

```
6   print("12: ",12)
7   print("13: ",13)
8   print("14: ",14)
9   >>>
10  11:   [1, 2, 3]
11  12:   [4, 5, 6]
12  13:   [1, 2, 3, 4, 5, 6]
13  14:   [1, 2, 3, 1, 2, 3, 1, 2, 3, 1, 2, 3]
```

The *in* operator may be used to test whether the list contains that element in which case the interpreter responds by *True* and *False* otherwise as shown below.

```
1   >>> l = [1, 'a', 3.45, 'hello']
2   >>> 'a' in l
3   True
4   >>> 5 in l
5   False
6   True
```

Copying a list to another should be done with care since equating the new list to the original one simple creates a reference and change in one of the lists affects the other. In order to create a new list, we can use the list method *copy* or simply use < *oldlist*[:] > as in the code below. Note that modifying second element in *l*1 affects *l*2 since they both are associated with the same list. However, using *copy* method or placing a colon in square brackets creates a new list which is not affected by any change in the original list as shown below.

```
1   l1 = [ 1, 2, 3, 4, 5]
2   l2 = l1
3   l3 = l1[:]
4   l4 = l1.copy()
5   l1[1] = 12
6   print("l1: ", l1)
7   print("l2: ",l2)
8   print("l3: ",l3)
9   print("l4: ", l4)
10  >>>
11  l1:   [1, 12, 3, 4, 5]
12  l2:   [1, 12, 3, 4, 5]
13  l3:   [1, 2, 3, 4, 5]
14  l4:   [1, 2, 3, 4, 5]
```

2.2.3 Tuples

Tuples are similar to lists except that they are immutable, that is, the values in a tuple may not be changed. This means operating on tuples is more efficient than lists, thus, if we know that we will not change values stored in a sequence, it will be convenient to use a tuple instead of a list. A tuple may be declared as a list but without square brackets or with no delimiters.

Most of the operators defined for lists are available for tuples. Conversion from a tuple to list may be done by the *list* method and a tuple may be converted to a list by the *tuple* method as in the following example where we declare a tuple *t* and attempt to modify its element with index 1 and get an error. When we convert this tuple to the list *tl*, we can modify the contents of the list. We then convert *tl* back to *t* using the *tuple* method and we can not modify its contents as before.

```
1   >>> t = (2, 9, 3, 5)
2   >>> t[0]
3   2
4   >>> t[1] = 5
5   Traceback (most recent call last):
6     File "<pyshell#84>", line 1, in <module>
7       t[1] = 5
8   TypeError: 'tuple' object does not support item assignment
9   >>> tl = list(t)
10  >>> tl[1] = 5
11  >>> tl
12  [2, 5, 3, 5]
13  >>> t = tuple(tl)
14  >>> t[2] = 8
15  Traceback (most recent call last):
16    File "<pyshell#89>", line 1, in <module>
17      t[2] = 8
18  TypeError: 'tuple' object does not support item assignment
```

2.2.4 Arrays

An array is like a list but consists of homogenous elements and can be multidimensional as lists. Arrays are stored more efficiently in memory and computations involving arrays are faster than list manipulations, thus, if all elements of a structure have the same type, it is beneficial to use arrays instead of lists. The main library to manage arrays in Python is *numpy* which needs to be imported prior to any computation. The following shows how to create two arrays *arr*1 and *arr*2 of dimension one with 5 elements and print them. The numpy *arange* method is used to form *arr*2 which simply assignes values 0 to 4 to *arr*2.

```
1   import numpy as np
2   arr1 = np.array([0, 1, 2, 3, 4])
3   arr2= np.arange(5)
4   print("arr1:",arr1)
5   print("arr2:",arr2)
6   >>>
7   arr1: [0 1 2 3 4]
8   arr2: [0 1 2 3 4]
```

An element of an array can be accessed and modified in the usual way using indices, for example, $arr1[3] = 5$ would set this entry to 3. A list can be converted to an array simply by using the *numpy* method *array* and an array can be converted to a list by the method *list* as in the example below. Note that display of a list elements has commas and array elements do not have commas in between them. An optional *dtype* parameter specifies the data type in an array when created.

```
import numpy as np
list1 = [1, 3, 5, 7]
arr = np.array(list1)
list2 = list(arr)
print("list1:",list1)
print("arr2:",arr)
print("list2:",list2)
>>>
list1: [1, 3, 5, 7]
arr2: [1 3 5 7]
list2: [1, 3, 5, 7]
```

We will be using the following methods in arrays frequently.

- *numpy.ones((m,n))*: Creates an array of all ones with dimension $m \times n$.
- *numpy.zeros((m,n))*: Creates an array of all zeros with dimension $m \times n$.
- *numpy.linspace(start,end,step)*: Creates an array starting with *start* end ending with *end* with space *step* between each element.
- *numpy.sort(array)*: Sorts array in ascending order. When the optional parameter [::−1] is specified, sorting is performed in descending order. Sorting a 2-D array results in sorting each row.
- *numpy.concatenate((a,b))*: Joins array b to the end of array a. .
- *numpy.copy(a)*: Copies array a to another one. This method is needed since simple array assignment such as $b = a$ just creates a reference to the first one resulting in modification in both when one is modified as described for lists.

The usual *max, min* methods for the lists are also available for arrays. We can create a boolean array by specifying the data type as boolean by the statement $dtype = bool$. Slicing an array using colons is possible as in the following example.

```
import numpy as np

arr1 = np.arange(10)
arr2 = np.array([[1,2,3,4],[5,6,7,8],[9,10,11,12]])
print(arr1[2:6]) # print elements from 2nd to 5th
print(arr1[:6])  # print elements up to 6th
print(arr1[6:])  # print elements after 6th
print(arr1[::2]) # print all with space 2
print(arr2)
print(arr2[1,:]) # print the 1st row of arr2
print(arr2[:,2]) # print the 2nd column of arr2
```

```
12   >>>
13   [2  3  4  5]
14   [0  1  2  3  4  5]
15   [6  7  8  9]
16   [0  2  4  6  8]
17   [[  1    2    3    4]
18    [  5    6    7    8]
19    [  9  10  11  12]]
20   [5  6  7  8]
21   [  3    7  11]
```

2.2.5 Dictionaries

A Python dictionary is a collection of unordered elements which consists of key and value pairs. A dictionary is enclosed within curly brackets containing *key : value* pairs with key being an immutable unique object. Values in a dictionary can be accessed and modified using key values. A dictionary may have numeric keys or string keys and the *pop* method removes the entry with the given key value from the dictionary. We can also use *del* method to remove an item if we do not want to capture the deleted item. The following code demonstrates these concepts.

```
1    student = {'name': 'Jose', 'number':1234}
2    print (student)
3    student['department'] = 'CS' # add department
4    print(student)
5    num = student.pop('number')            # remove number
6    print(num)
7    print(student)
8    student['department'] = 'MATH' # update department
9    print(student)
10   >>>
11   {'name': 'Jose', 'number': 1234}
12   {'name': 'Jose', 'number': 1234, 'department': 'CS'}
13   1234
14   {'name': 'Jose', 'department': 'CS'}
15   {'name': 'Jose', 'department': 'MATH'}
```

The *key* method as applied to a dictionary lists all of the keys and *values* method lists all values stored in the dictionary as shown below.

```
1    dict1 = {'type':'book', 'pages':120, 'price': 34.9}
2    print("keys: ",dict1.keys())
3    print("values: ",dict1.values())
4    >>>
5    keys:  dict_keys(['type', 'pages', 'price'])
6    values:  dict_values(['book', 120, 34.9])
```

2.2.6 Sets

A Python set is an unordered collection of mutable elements that have no duplicates. An element can be added to a set by the *add* method, multiple elements can be inserted by the *update* method and removing an element is done by the *remove* method. All of these methods are demonstrated in the following code.

```
set1 = set(['a','b','c','a','bc','ab','aa','b'])
print(set1)
set1.add('cc')   # add 'cc' to set
print(set1)
set1.update([12,23]) # update set
print(set1)
set1.remove('ab')    # remove 'ab' from set
print(set1)
>>>
{'ab', 'bc', 'a', 'aa', 'b', 'c'}
{'ab', 'bc', 'a', 'aa', 'b', 'cc', 'c'}
{'ab', 'bc', 12, 'a', 'aa', 'b', 23, 'cc', 'c'}
{'bc', 12, 'a', 'aa', 'b', 23, 'cc', 'c'}
```

2.3 Flow Control

Control of flow in a Python program may be accomplished by decision control where the flow may divert to one of few possible directions, or loops where an operation is repeated possibly with different data.

2.3.1 The *if* Statement

As common in various other languages, the logical statement immediately after *if* is checked and the next statement is executed if this logical statement yields a *true* value. The *else if* (*elif*) and *else* statements function as to test alternatives when the statement after *if* is *false*. The following example code inputs an integer from the keyboard and determines whether this input is positive, zero or negative.

```
a = int(input("Please enter an integer: "))
if a < 0:
    print('Negative')
elif a == 0:
    print('Zero')
else:
    print('Positive')
```

The if statements may be nested in which case *else* and *elif* blocks refer to the last *if* statement as shown in the code below where we test whether an input number is between 3 and 12.

```
a = eval(input("enter a number "))
if a > 3:
    if a < 12:
        print("between 3 and 12")
    else:
        print("greater than 12")
else:
    print("less than 3")
>>> enter a number 6
between 3 and 12
>>> enter a number 23
greater than 12
>>> enter a number -5
less than 3
```

2.3.2 Loops

The main loop structures we can use in Python are the *while* and *for* loops.

2.3.2.1 The *while* Loop

The *while* loop executes if the logical statement after *while* yields a true value in the usual sense. The following Python code inputs integers until a negative integer is input and prints the sum of the integers entered. Note that we need two input statements, the first one for the *while* loop to work for the first time and the second one to iterate inside the loop. Also, we used *int* method to convert the input string as an alternative to the *eval* method.

```
s = 0
a = int(input("Please enter an integer: "))
while a >= 0:
    s = s + a
    a = int(input("Please enter an integer: "))
print('sum of entered integers:', s)
>>>
Please enter an integer: 3
Please enter an integer: 6
Please enter an integer: 2
Please enter an integer: -1
sum of entered integers: 11
```

2.3.2.2 The *for* Loop

The *for* loop structure in Python is slightly different than C or Pascal implementations. There are few different implementations of a loop using the *for* statement; we can iterate through a set of values in sequence which is similar to *for all* statement in pseudocode or specify a range of values for the *for* loop index as shown in the code below. The list *morning* contains three words of different lengths and the first *for* loop iterates through elements of this list and prints each word. The second loop iterates through the index values of the list using the *range* function which assigns values $0, \ldots,$ (length of the list) -1 value to the index.

```
1   morning = ['good', 'morning ', 'all']
2
3   for w in morning:
4       print(w, len(w))
5   for i in range(len(morning)):
6       print(morning[i],len(morning[i]))
7   >>>
8   good 4
9   morning  8
10  all 3
11  good 4
12  morning  8
13  all 3
```

The *break* and *continue* statements inside these loops work similarly as in other languages, the *break* statement exits the loop and *continue* statement continues with the next iteration ignoring the statements between *continue* and the end of the loop. The following example shows how to stop and skip the letter *o* using these statements in the word "Algorithm".

```
1   for w in 'Algorithm':
2       if w == 'o':
3           break
4       print(w, end =" ")
5   print("\n")
6
7   for w in 'Algorithm':
8       if w == 'o':
9           continue
10      print(w,end =" ")
11  >>>
12  A l g
13
14  A l g r i t h m
```

List comprehension is a method to create a list, commonly using a *for* loop. The following examples show how to create a list of ordered integers, their squares and creating a list using another list. If the source list is the same as the target list, then the list elements are modified accordingly.

```
1   L0 = [ 1, 5, 3, 9, 7, 14]
2   print("L0:", L0)
3   L1 = [i for i in range(1,10)]
4   L2 = [i**2 for i in range(1,10)]
5   L3 = [i*2  for i in L0]
6   L4 = [i  for i in L0 if i >= 5]
7   L0 = [i+1  for i in L0]
8
9   print("L1:", L1)
10  print("L2:", L2)
11  print("L3:", L3)
12  print("L4:", L4)
13  print("L0 modified:", L0)
14  >>>
15  L0: [1, 5, 3, 9, 7, 14]
16  L1: [1, 2, 3, 4, 5, 6, 7, 8, 9]
17  L2: [1, 4, 9, 16, 25, 36, 49, 64, 81]
18  L3: [2, 10, 6, 18, 14, 28]
19  L4: [5, 9, 7, 14]
20  L0 modified: [2, 6, 4, 10, 8, 15]
```

2.4 Functions

A function in Python is a subprogram that may be called from anywhere in the program as in various other languages. Functions are defined by the *def* keyword followed by the name of the function and optional parameters passed to the function. A function returns a value by the *return* statement and if this statement is missing, the returned value is *None* which means nothing is returned.

The following example illustrates how to pass an integer to a function named *find_sum* which calculates the sum of integers up to and including that integer and returns this value.

```
1   def find_sum(n):
2       s = 0
3       for i in range(1,n+1):
4           s = s + i
5       return s
6
7   for i in range(1,11):
8       print(find_sum(i), end=" ")
9   >>>
10  1 3 6 10 15 21 28 36 45 55
```

A function may return more than one value as shown in the following example where we use the *numpy sort* method.

```
1
2   import numpy as np
3
4   def max_sort(a):
5       m = max(a)
6       b = np.sort(a)
7       return m,b
8
9   arr = np.array([5,1,8,3,0])
10  m,c = max_sort(arr)
11  print(m,c)
12  >>>
13  8 [0 1 3 5 8]
```

Functions can be defined without using formal function definition using the *lambda* keyword as in the following example where we define the *cube* function in one line.

```
1   >>> cube = lambda x:x**3
2   >>> cube(5)
3   125
```

Scope of Variables

A variable defined inside a function has local significance only. A variable defined external to a function is global and may be read but not modified by a function, it should be declared global in the function to be modified. Arguments passed to a function may be in arbitrary order when the name of the parameter is specified upon calling the function. These concepts are illustrated in the following code where the number of occurrences of a letter in a word is counted and returned along with an incremented value of an input variable.

```
1   word = 'Abracadabra'
2
3   def count(c, x):
4       n = 0
5       for w in word:
6           if w == c:
7               n = n + 1
8       x = x + 1
9       return n, x
10
11  print(count(x=5, c='a'))
12  >>>
13  (4, 6)
```

2.5 Modules

A *module* is a collection of Python programs that perform related tasks. Consider
for example the need to group various queue operations such as *enqueue*, *dequeue*
etc. into a module called *queue*. The simplest way to use the functions of this module
is to import the module before using the functions as below.

```
import queue as q
q.enqueue(q1,data1)
```

Using the abbreviation for the module name is optional, we could have used just
import queue and then use the full name of the module everytime we need to invoke
a method from this module. It is possible to import only the names of the methods
from the module as below.

```
from queue import enqueue
enqueue(q1,data1)
```

The name of a module is the same as the name of the file that stores the module
code. Let us write a module named *arrayop* that has functions to find the average
value stored in an array which contains numbers and another function that creates
an array which is the reverse of its input array as below.

```
import numpy as np

def find_ave(a):
    n = len(a)
    s = 0
    for i in range(n):
        s = s + a[i]
    ave = s / n
    return ave

def reverse_arr(a):
    n = len(a)
    b = np.zeros((n))
    for i in range(n):
        b[i] = a[n-i-1]
    return b
```

We can now import this module and use the functions inside the module as below.

```
import numpy as np
import arrayop as ap

arr = np.array([3,1,4,5])
print(ap.find_ave(arr))
print(ap.reverse_arr(arr))
>>>
```

```
8    3.25
9    [5. 4. 1. 3.]
```

The module *time* contains the method *time* to measure the execution time between two points of a program as shown below. We create an array of 10^5 elements with each element having the same value of its index. We then record time to find the sum of array elements using a *for* loop first and then by using the *sum* method of arrays. The *sum* method results in about half of the time (0.026 s) of the first method which spent 0.045 s.

```
1    import numpy as np
2    import time
3
4    a = np.arange((100000))
5    s1 = 0
6    s2 = 0
7    start = time.time()
8    for i in range(100000):
9        s1 = s1 + a[i]
10   stop = time.time()
11   print("time by loop:", stop-start, "sum:", s1)
12   start = time.time()
13   s2 = sum(a)
14   stop = time.time()
15   print("time using sum:", stop-start, "sum:", s2)
16   >>>
17   time by loop: 0.04534482955932617 sum: 4999950000
18   time using sum: 0.02639293670654297 sum: 4999950000
```

Some commonly used modules are the *sys* module and the *os* module for operating system related functions. A *package* is a directory that can contain more than one module. It should contain a special file called *__init__.py* which tells Python interpreter that the directory is a package. Consider a package called *project* that has a module *mod.py*. Invoking a function $f(x)$ that finds the square of its input from module *mod* may be done by the following.

```
1    import project
2    a = project.mod.f(5)
3    print(a)
4    >>>
5    25
```

2.6 Chapter Notes

We reviewed basic Python features that will aid the design of algebraic graph algorithms in this chapter. The emphasis was in properties of Python we need to use when dealing with algorithms rather than a general introduction to this language. We will often get involved in declaring arrays, and manipulation of them such as finding maximum/minimum value in an array, sorting an array etc. In various cases, lists will be our attention when we need to do operations such as appending an element to a list or extending a list by another one. Our review is only very basic and confined to what we need as stated. Detailed reviews and tutorials [1] can be found on the Web.

Programming Exercises

1. Write a single Python statement that finds the count of integers greater than 2 in a given list using list comprehension.
2. A list L containing integers between 0 and 50 is given. Write a single statement that finds the frequencies of integers in this list and stores these values in another list using list comprehension.
3. Write a Python program that generates random numbers between 0 and 1 for 10 times and appends each generated number to a list which is initialized as empty.
4. Write a Python program that inputs integers until a 0 is encountered and then prints the maximum value entered up to that point.
5. Write a Python function that inputs an array of integers and returns the maximum value in this array. Do not use the max method on array class.
6. Write a Python program that inputs a sentence string by the user and counts the number of words in the sentence.
7. Write a Python program that inputs a large integer and outputs this integer with commas inserted at every 3 digits. For example, when 5231908126 is entered, the output is 5,231,908,126.

Reference

1. https://docs.python.org/tutorial.pdf

Basic Matrix Computations

3

Abstract

In this chapter, we review basic matrix theory and matrix algebra with algorithms to perform matrix operations. We first define matrices, describe operations on matrices and review basic matrix types. Matrix multiplications are basic operations we will use in various algebraic graph algorithms and we describe algorithms for this purpose in detail. We conclude with matrix determinant, matrix inverse and matrix eigenvalues and eigenvectors. We implement most of the algorithms with available functions from Python libraries.

3.1 Introduction

An $m \times n$ matrix contains $m \cdot n$ real numbers arranged in m rows and n columns. The element at ith row and jth column of a matrix A is denoted by a_{ij} as shown below.

$$\begin{bmatrix} a_{11} & a_{12} & \dots & a_{1n} \\ a_{21} & a_{22} & \dots & a_{2n} \\ \vdots & \vdots & \ddots & \vdots \\ a_{m1} & a_{m2} & \dots & a_{mn} \end{bmatrix}$$

A *square matrix* has the same number of rows and columns. An $1 \times n$ matrix is called a *row vector* and an $m \times 1$ matrix is called *column vector* as follows.

$$\begin{bmatrix} a_1 & a_2 & a_3 & \cdots & a_n \end{bmatrix}, \quad \begin{bmatrix} a_1 \\ a_2 \\ \dots \\ a_m \end{bmatrix}$$

K. Erciyes, *Algebraic Graph Algorithms*, Undergraduate Topics in Computer Science, https://doi.org/10.1007/978-3-030-87886-3_3

3.2 Matrix Operations

We will review basic matrix operations in this section. Let A and B be two matrices of size $m \times n$. The *sum* of A and B, $A + B$, is the matrix formed by adding the corresponding elements of A and B.

$$A + B = \begin{bmatrix} a_{11} + b_{11} & a_{12} + b_{12} & \cdots & a_{1n} + b_{1n} \\ a_{21} + b_{21} & a_{22} + b_{22} & \cdots & a_{2n} + b_{2n} \\ \cdots & \cdots & \cdots & \cdots \\ a_{m1} + b_{m1} & a_{m2} + b_{m2} & \cdots & a_{mn} + b_{mn} \end{bmatrix}$$

Example 3.2.1 Let A and B be two matrices given below. Find $A + B$.

$$A = \begin{bmatrix} 3 & 1 & 7 \\ 1 & 5 & 4 \\ 6 & 4 & 2 \end{bmatrix}, \qquad B = \begin{bmatrix} 1 & 4 & 2 \\ 6 & 5 & 1 \\ 2 & 1 & 2 \end{bmatrix}$$

The sum of A and B is as below,

$$A + B = \begin{bmatrix} 4 & 5 & 9 \\ 7 & 10 & 5 \\ 8 & 5 & 4 \end{bmatrix}$$

Similarly, $A - B$ is obtained by subtracting each element of B from the corresponding element of B. The *product* of matrix A by a scalar k, $k \cdot A$ or kA, is the matrix formed by multiplying each element of A by k as below.

$$kA = \begin{bmatrix} ka_{11} & ka_{12} & \cdots & ka_{1n} \\ ka_{21} & ka_{22} & \cdots & ka_{2n} \\ \cdots & \cdots & \cdots & \cdots \\ ka_{m1} & ka_{m2} & \cdots & ka_{mn} \end{bmatrix}$$

For the matrix A in Example 3.2.1,

$$5A = \begin{bmatrix} (5 \cdot 3) & (5 \cdot 1) & (5 \cdot 7) \\ (5 \cdot 1) & (5 \cdot 5) & (5 \cdot 4) \\ (5 \cdot 6) & (5 \cdot 4) & (5 \cdot 2) \end{bmatrix} = \begin{bmatrix} 15 & 5 & 35 \\ 5 & 25 & 20 \\ 30 & 20 & 10 \end{bmatrix}$$

Let A, B and C be three matrices with the same dimensions and k be a scalar. Then, the following properties of matrix addition and multiplication by a scalar can be stated.

1. $A + B = B + A$ 3. $kA = Ak$
2. $(A + B) + C = A + (B + C)$ 4. $k(A + B) = kA + kB$

The library associated with basic matrix operations in Python is called *numpy* (Numeric Python) and it needs to be imported before any matrix operation. A matrix is formed and initialized by the library function *numpy.array([[....], [...],[...]])*. In the following, an array named A is declared with the given values and it is printed.

```
1    import numpy as np
2    A = np.array([[1,2],[3,4]])
3    print (A)
4    >>>
5    [[1 2]
6     [3 4]
```

Python *numpy* library functions to perform basic matrix operations are as follows.

- *C=numpy.add(A, B)*: Add two matrices *A* and *B* and store result in *C*.
- *C=numpy.subtract(A, B)*: Subtract matrix *B* from matrix *A* and store result in *C*.
- *C=numpy.divide(A, B)*: Divide matrix *A* by matrix *B* and store result in *C*.
- *C=numpy.multiply(A, B)*: Multiply matrix *A* by matrix *B* and store result in *C*.
- *C=numpy.sum(A)*: Form the sum of elements of matrix *A* and store result in $c \in \mathbb{R}$.
- *C=numpy.sum(A, axis = 0)*: Form the column wise summation of matrix *A* and store result in vector *C*.
- *C=numpy.sum(A, axis = 1)*: Form the row wise summation of matrix *A*, store result in vector *C*.

The following Python code shows the implementation of these methods in a sample matrix.

```
import numpy as np

A = np.array([[1,2,3],[4,5,6]])
B = np.array([[7,8,9],[10,11,12]])
C = np.add(A,B)
print("Sum:\n",C)
C = np.subtract(A,B)
print("Difference:\n",C)
C = np.multiply(A,A)
print("Product:\n",C)
C = np.divide(A,B)
print("Quotient:\n",C)
s = np.sum(A)
print("Sum of A:\n",s)
C = np.sum(A,axis=1)
print("Sum vector of rows of A:\n",C)
C = np.sum(A,axis=0)
print("Sum vector of columns of A:\n",C)
>>>
Sum:
 [[ 8 10 12]
 [14 16 18]]
Difference:
 [[-6 -6 -6]
 [-6 -6 -6]]
Product:
```

```
[[  1   4   9]
 [16 25 36]]
Quotient:
 [[0.14285714 0.25        0.33333333]
 [0.4         0.45454545 0.5         ]]
Sum of A:
 21
Sum vector of rows of A:
 [ 6 15]
Sum vector of columns of A:
 [5 7 9]
```

3.3 Transpose of a Matrix

The transpose of a matrix A denoted by A^T or A' is obtained by converting rows of A to columns in order as shown below.

$$A = \begin{bmatrix} 2 & 1 & 4 \\ 3 & 6 & 5 \end{bmatrix}, \qquad A^T = \begin{bmatrix} 2 & 3 \\ 1 & 6 \\ 4 & 5 \end{bmatrix}$$

The transpose of a row vector is a column vector and likewise, transpose of a column vector is a row vector. The following properties of the transpose operation can be verified for matrices A and B.

- $(A^T)^T = A$
- $(A + B)^T = A^T + B^T$
- $(AB)^T = B^T A^T$

The transpose of a matrix in Python is obtained by calling the method *transpose* or simply T on an *numpy* array as in the code below.

```
1   import numpy as np
2
3   A = np.array([[2,1,4],[3,6,5]])
4   B = A.T
5   print(B)
6   >>>
7   [[2 3]
8    [1 6]
9    [4 5]]
```

3.4 Matrix Properties

Definition 3.1 *(symmetric matrix)* A matrix A is called *symmetric* if $A^T = A$. In other words, $\forall i, j; a_{ij} = a_{ji}$.

Example 3.4.1 The following matrix is syymmetric.

$$\begin{bmatrix} 2 & 1 & 5 \\ 1 & 4 & 3 \\ 5 & 3 & 7 \end{bmatrix}$$

Definition 3.2 *(antisymmetric matrix)* A matrix A is called *antisymmetric* if $A^T = -A$. In other words, $\forall i, j; a_{ij} = -a_{ji}$.

Example 3.4.2 The following matrix is antisymmetric. Note that we have 0s in the diagonal of this matrix to have $a_{ii} = -a_{ii}$.

$$\begin{bmatrix} 0 & 3 & -4 \\ -3 & 0 & 1 \\ 4 & -1 & 0 \end{bmatrix}$$

Definition 3.3 *(orthogonal matrix)* A matrix A is called orthogonal if $A^T A = AA^T = I$.

Example 3.4.3 The following matrix is orthogonal. Note that we have 0s in the diagonal of this matrix to have $a_{ii} = -a_{ii}$.

$$\begin{bmatrix} 0 & 3 & -4 \\ -3 & 0 & 1 \\ 4 & -1 & 0 \end{bmatrix}$$

3.5 Types of Matrices

We will review special matrices in this section starting with the identity matrix.

3.5.1 Diagonal Matrix

The diagonal of a matrix A has elements with the same subscripts such as a_{11}, a_{22}, a_{33} etc. A *diagonal matrix* has non-zero elements only in its diagonal. The following matrix is a diagonal matrix.

$$\begin{bmatrix} 3 & 0 & 0 \\ 0 & -1 & 0 \\ 0 & 0 & 2 \end{bmatrix}$$

Definition 3.4 *(trace of a matrix)* The trace of a matrix is the sum of its diagonal elements.

Trace of a matrix is equal to the transpose of its trace. The elements along the diagonal of a square matrix may be extracted by the *diag* method of *numpy*. The optional parameter k provides retrieving values on diagonals above or below the main diagonal. Trace of a matrix may be obtained by the call *trace* as in the code below.

```
import numpy as np

A = np.array([[1,2,3],
              [4,5,6],
              [7,8,9]])
print("diagonal:",np.diag(A))
print("upper diagonal:",np.diag(A,k=1))
print("lower diagonal",np.diag(A,k=-1))
print("trace:",np.trace(A))
>>>
diagonal: [1 5 9]
upper diagonal: [2 6]
lower diagonal [4 8]
trace: 15
```

3.5.2 Identity Matrix

The *identity matrix* I is a square matrix with all 1's in its diagonal. The following is a 3×3 identity matrix. Any matrix multiplied by the identity matrix is equal to itself. That is, $AI = IA = A$ for any square matrix. A 3×3 identity matrix is as follows.

$$\begin{bmatrix} 1 & 0 & 0 \\ 0 & 1 & 0 \\ 0 & 0 & 1 \end{bmatrix}$$

The trace of $n \times n$ identity matrix is equal to n. The following properties of operations of matrix $A_{m,n}$ with the identity matrix I are evident.

- $AI = IA = A$
- $(kI_m)A = k(I_m A) = kA$
- $AI_n = I_m A = A$

3.5.3 Triangular Matrix

A matrix is called an *upper triangular matrix* if all of its elements below the diagonal are zero and it is called a *lower triangular matrix* if all of its elements above the diagonal are zero. The call *triu* from *numpy* library provides forming an upper triangular matrix and *tril* forms a lower triangular matrix as shown below.

```
import numpy as np

A = np.array([[1,2,3],
              [4,5,6],
              [7,8,9]])
print("upper:\n",np.triu(A))
print("lower:\n",np.tril(A))
>>>
upper:
 [[1 2 3]
 [0 5 6]
 [0 0 9]]
lower:
 [[1 0 0]
 [4 5 0]
 [7 8 9]]
```

3.6 Matrix Multiplication

Matrix multiplication is one of the most common operations performed by matrices. We will review basic vector-vector, matrix-vector and matrix-matrix multiplication in this section.

3.6.1 Vector-Vector Multiplication

The product of a row vector A with a column vector B is obtained by multiplying the corresponding entries and adding as below. Note that the size of each vector should be equal. This operation results in a single element vector as shown in the code below.

$$AB = \begin{bmatrix} a_1 & a_2 & a_3 & \cdots & a_n \end{bmatrix} \begin{bmatrix} a_1 \\ a_2 \\ \cdots \\ a_n \end{bmatrix} = a_1b_1 + a_2b_2 + \cdots + a_nb_n = \sum_{k=1}^{n} a_kb_k$$

Example 3.6.1

$$AB = \begin{bmatrix} 3 & 1 & 0 & 5 & 2 \end{bmatrix} \begin{bmatrix} 4 \\ 2 \\ 7 \\ 3 \\ 6 \end{bmatrix} = 3 \cdot 4 + 1 \cdot 2 + 0 \cdot 7 + 5 \cdot 3 + 2 \cdot 6 = 41$$

A vector-vector multiplication algorithm is shown in Algorithm 3.1 where x and y are two vectors of size n and the resulting value is stored in c. This function has $\Theta(n)$ complexity as done in the *for* loop.

Algorithm 3.1 *Matrix-Matrix Multiplication*

1: **Input** : vector x_n, vector y_n
2: **Output** : product $c \in \mathbb{R}$
3: $c \leftarrow \emptyset$
4: **for** $j = 1$ to n **do**
5:     $c = c + x_i \cdot y_i$
6: **end for**

Two arrays are declared by the Python *numpy* library method *array* and the multiplication is done using the *dot* function in the code below. Running this code section results in the output 32.

```
import numpy as np

A = np.array([1,2,3])
B = np.array([4,5,6])
C = np.dot(A,B)
print(C)
>>>
32
```

3.6.2   Matrix-Vector Multiplication

Matrix vector multiplication can be envisioned as multiple vector-vector multiplication when each row of a matrix is considered as a vector. An example multiplication of a 3×4 matrix by a vector of 4 elements is shown below. Note that the product is a vector of 4 elements with element i corresponding to the product of the row i of matrix with the vector.

$$\begin{bmatrix} 1 & 4 & 2 & 0 \\ 2 & 0 & 1 & 5 \\ 7 & 1 & 6 & 3 \end{bmatrix} \begin{bmatrix} 3 \\ 1 \\ 0 \\ 4 \end{bmatrix} = \begin{bmatrix} 7 \\ 26 \\ 34 \end{bmatrix}$$

The algorithm to perform this multiplication consists of two nested *for* loops with the outer loop indexed by the rows of the matrix and the inner loop for multiplying each jth element of the ith row of matrix with the jth element of the vector. The time complexity of this algorithm is $O(n^2)$ due to two nested loops.

Algorithm 3.2 *Matrix-Vector Multiplication*

1: **Input** : matrix $A_{m \times n}$, vector x_n
2: **Output** : product $c \in \mathbb{R}$
3: $y \leftarrow \emptyset$
4: **for** $i = 1$ to m **do**
5:     **for** $j = 1$ to n **do**
6:         $y_i = y_i + a_{ij} \cdot x_j$
7:     **end for**
8: **end for**

Python code to implement this operation is shown below.

```
import numpy as np

a = np.array([[1,1,1]])
b = np.array([[1,2,3],[4,5,6],[7,8,9]])
c = np.dot(a,b)
print(c)
>>>
[[12 15 18]]
```

3.6.3 Matrix-Matrix Multiplication

Let A and B be two matrices such that the number of columns k of A is equal to the number of rows k of B. The product $C = AB$ is the $m \times n$ matrix with entry c_{ij} formed by vector multiplication of the ith row of A with the jth column of B.

Example 3.6.2 Find AB with,

$$A = \begin{bmatrix} 4 & 1 & 5 \\ 2 & 0 & 1 \end{bmatrix}, \qquad B = \begin{bmatrix} 2 & 1 \\ 3 & 5 \\ 1 & 7 \end{bmatrix}$$

The matrix A has 2 rows and B has two columns, thus, we can multiply them as below. The product matrix C is 2×2.

$$C = AB = \begin{bmatrix} (4 \cdot 2) + (1 \cdot 3) + (5 \cdot 1) & (4 \cdot 1) + (1 \cdot 5) + (5 \cdot 7) \\ (2 \cdot 2) + (0 \cdot 3) + (1 \cdot 1) & (2 \cdot 1) + (0 \cdot 5) + (1 \cdot 7) \end{bmatrix} = \begin{bmatrix} 16 & 43 \\ 5 & 9 \end{bmatrix}$$

Let A, B and C be three matrices with the same dimensions and k be a scalar. Then, the following properties of matrix multiplication can be stated.

1. $A(BC) = (AB)C$
2. $(A + B)C = AC + BC$
3. $A(B + C) = AB + AC$
4. $k(AB) = (kA)B = A(kB)$

Multiplication of an $m \times n$ matrix A by another $n \times r$ matrix B is performed in the usual way as multiplying the ith row of A with the jth column of B to form c_{ij} entry of the product matrix C. A matrix-matrix multiplication based on this method can be formed as in Algorithm 3.2. Again, this multiplication may be viewed as vector-vector product of ith row of matrix A with the jth column of matrix B.

Algorithm 3.3 *Matrix-Matrix Multiplication*

1: **Input** : matrix $A_{m \times r}$ and matrix $B_{r \times n}$
2: **Output** : matrix $C_{m \times n}$
3: $C \leftarrow \emptyset$
4: **for** $i = 1$ to m **do**
5:     **for** $j = 1$ to n **do**
6:         $c_{ik} \leftarrow 0$
7:         **for** $k = 1$ to r **do**
8:             $c_{ik} = c_{ik} + a_{ij} \cdot b_{jk}$
9:         **end for**
10:     **end for**
11: **end for**

Clearly, this naive algorithm takes $O(nmp)$ operations or $O(n^3)$ when both input matrices are square with dimension n. An example multiplication of a 3×4 matrix by a 4×2 matrix to yield a 3×2 matrix is shown below.

$$\begin{bmatrix} 2 & 1 & 0 & 3 \\ 1 & 0 & 3 & 2 \\ 6 & 1 & 2 & 3 \end{bmatrix} \begin{bmatrix} 1 & 4 \\ 2 & 0 \\ 7 & 1 \\ 0 & 3 \end{bmatrix} = \begin{bmatrix} 4 & 17 \\ 22 & 13 \\ 22 & 35 \end{bmatrix}$$

Multiplication of these two matrices is performed again by the *dot* method as in the example code below.

```
import numpy as np

A = np.array([[2,1,0,3],[1,0,3,2],[6,1,2,3]])
B = np.array([[1,4],[2,0],[7,1],[0,3]])
C = np.dot(A,B)
print(C)
>>>
[[ 4 17]
 [22 13]
 [22 35]]
```

3.6.4 Boolean Matrix Multiplication

A Boolean matrix A has elements $a_{ij} \in \{True, False\}$. The Boolean matrix-matrix multiplication is performed similar to the ordinary matrix-matrix multiplication but we need to replace lines 7–9 of Algorithm 3.3 with the following,

$$c_{ij} = \bigvee_{k=1}^{n} a_{ik} \wedge b_{kj}$$

This multiplication requires $O(n^3)$ operations using the naive algorithm as with the ordinary multiplication.

Definition 3.5 *(witness matrix)* Consider the matrix multiplication $C = AB$. A matrix W is called a *witness matrix* with the following conditions:

- If $c_{ij} = 0$, then $w_{ij} = 0$.
- If $c_{ij} = 1$, then $w_{ij} = k$ where $a_{ik} = b_{kj} = 1$.

The following code displays two Booelan matrices A and B and their product C. Note that we define these matrices with 0 and 1 entries and declare their types as boolean with $dtype = bool$ statement after specifying the entries. Alternatively, we could have specified theses matrices with *True* and *False* entries. The boolean product C is then converted to integer by the *astype* method and both forms of C are printed as shown.

```
import numpy as np

A = np.array([[1,0,1],[0,1,0],[0,0,1]],dtype=bool)
B = np.array([[1,0,1],[0,1,0],[0,0,1]],dtype=bool)
C = np.dot(A,B)
print(C)
print(C.astype(int))
>>>
[[ True False  True]
 [False  True False]
 [False False  True]]
[[1 0 1]
 [0 1 0]
 [0 0 1]]
```

3.7 Determinant of a Matrix

Consider the 2×2 matrix A given below.

$$\begin{bmatrix} a_{11} & a_{12} \\ a_{21} & a_{22} \end{bmatrix}$$

The determinant of A is defined by,

$$\det A = a_{11}a_{22} - a_{12}a_{21}$$

The Python method *det* from *numpy linalg* library can be used to find the determinant of a square matrix as shown below.

```
import numpy as np
A = np.array([[1,2],[3,4]])
np.linalg.det(A)
-2.0
```

Computing determinant of a square matrix with larger dimension than 2 may be realized by reducing the determinant to that of dimension 2.

Example 3.7.1

$$\begin{bmatrix} a_{11} & a_{12} & a_{13} \\ a_{21} & a_{22} & a_{23} \\ a_{31} & a_{32} & a_{33} \end{bmatrix}$$

$$\det A = a_{11} \begin{vmatrix} a_{22} & a_{23} \\ a_{32} & a_{33} \end{vmatrix} - a_{12} \begin{vmatrix} a_{21} & a_{23} \\ a_{31} & a_{33} \end{vmatrix} + a_{13} \begin{vmatrix} a_{21} & a_{22} \\ a_{31} & a_{32} \end{vmatrix}$$

Example 3.7.2 Find the determinant of the square matrix A given below.

$$\begin{bmatrix} 2 & 1 & 3 \\ 5 & 4 & -1 \\ 3 & 2 & 6 \end{bmatrix}$$

$$\det A = 2 \cdot \begin{vmatrix} 4 & -1 \\ 2 & 6 \end{vmatrix} - 1 \cdot \begin{vmatrix} 5 & -1 \\ 3 & 6 \end{vmatrix} + 3 \cdot \begin{vmatrix} 5 & 4 \\ 3 & 2 \end{vmatrix} = 2(24+2) - 1(30+3) + 3(10-12) = 13$$

Let us check this result with the following Python code.

```
import numpy as np

A = np.array([[2,1,3],[5,4,-1],[3,2,6]])
d = np.linalg.det(A)
print(round(d,2))
>>>
13.0
```

3.8 Matrix Inverse

A square matrix A has an inverse if there exists a matrix B such that,

$$AB = BA = I$$

The matrix B denoted by A^{-1} is called the *inverse* of matrix A and it is unique. The defined inverse is called the right inverse since we have multiplied A by B on the right. The left inverse of A is defined as the matrix A^{-1} such that,

$$A^{-1}A = I$$

However, since he left inverse of a matrix is equal to its right inverse, we will use the term inverse without left or right orientation.

Theorem 3.1 *A matrix is invertible if and only if* $det(A)\, c \neq 0$. *In such a case, A is called a non-singular matrix; otherwise it is singular.*

Inverse of a matrix can be computed using the *numpy.linalg* library with the system call *inv*. The following Python code computes the inverse of a simple 2×2 matrix A. We then multiply the inverse of A with itself to give the identity matrix I for verification.

```
import numpy as np

A = np.array([[1,2],[3,4]],dtype=int)
B = np.linalg.inv(A)
print("A:\n",A)
print("B:\n",B)
C = np.dot(A,B).astype(int)
print("C:\n",C)
>>>
A:
 [[1 2]
 [3 4]]
B:
 [[-2.   1. ]
 [ 1.5 -0.5]]
C:
 [[1 0]
 [0 0]]
```

3.9 Rank Computation

A column of a given matrix is *dependent* on other columns if the values contained in that column can be generated by a weighted sum of any combination of other columns. A dependent row of a matrix is defined similarly for rows.

Definition 3.6 *(rank)* The rank of a matrix is the number of independent rows or columns it has.

The rank of a matrix may be found by the *rank* method from Python *numpy linalg* library. The following code demonstrates how rank of a 3×3 matrix with 2 independent rows is computed. Note that elements in row 2 of this matrix is twice of the values stored in the 0th row, thus, its rank is 2.

```
import numpy as np

A = np.array([[1,2,3],[4,5,6],[2,4,6]])
print("Rank of A:",np.linalg.matrix_rank(A))
>>>
Rank of A: 2
```

3.9.1 Powers of a Matrix

Computing powers of a matrix has may be used in various applications such as cryptography. The *n*th power of a matrix A is denoted by A^n in the usual manner and is equal to A multiplied n times by itself. A simple algorithm to find *n*th power of matrix A is shown in Algorithm 3.4. We will call this algorithm *iterative power* since it iteratively computes the power of a matrix by successively multiplying the product by the matrix.

Algorithm 3.4 *Iterative Matrix Power*

1: **procedure** POWER_ITER(A: $n \times n$ matrix, k: integer)
2:     $P = A$
3:     **for** $i = 1$ to $n - 1$ **do**
4:         $P = P \cdot A$
5:     **end for**
6:     **return** P
7: **end procedure**

The Python iterative matrix power algorithm that finds the 4th power of a sample matrix is shown below.

```
#################################################################
#                 Iterative Matrix Power                       #
#################################################################

import numpy as np

def Pow_Iter(A,k):
    n=len(A)
```

```
9       P = A
10      for i in range(1,k):
11          P = np.dot(P,A)
12      return P
13
14  if __name__ == '__main__':
15      B = np.array([[1,2,3],
16                    [1,1,1],
17                    [2,2,2]])
18      print(Pow_Iter(B,4))
19  >>>
20  [[209 250 291]
21   [104 125 146]
22   [208 250 292]]
```

Let us consider k as an even integer, then $A^k = A^{k/2} \cdot A^{k/2} = A^{k/4} \cdot A^{k/4} c \cdot A^{k/2} \cdot A^{k/4}$ and so on until A is reached. For example, $A^8 = A^4 \cdot A^4 = A^2 \cdot A^2 \cdot A^2 \cdot A^2 = A \cdot A \cdot A \cdot A \cdot A \cdot A \cdot A \cdot A$. When k is an odd integer, we need to start dividing by 2 from $k - 1$ and multiply the result by k. For example, $A^5 = A^4 \cdot A = A^2 \cdot A^2 \cdot A = A \cdot A \cdot A \cdot A \cdot A$. A recursive power algorithm may be designed with this observation which divides the power integer into 2 at each recursive call as in Algorithm 3.5.

Algorithm 3.5 *Recursive Matrix Power*

1: **procedure** POWER(A, k)
2: **if** $n == 1$: **return** A
3:     **if** $k\%2 = 0$ **then**
4:         **return** $Power(A, k/2) \cdot Power(A, k/2)$
5:     **else**
6:         **return** $A \cdot Power(A, (k - 1)/2) \cdot Power(A, (k - 1)/2)$
7:     **end if**
8: **end procedure**

We now find the 5th power of the same matrix using the recursive algorithm as in the Python code below. The base case of recursion is when $k = 1$ at which point we return A, otherwise, this function is called recursively.

```
1   ##############################################################
2   #                  Recursive Matrix Power                    #
3   ##############################################################
4
5   import numpy as np
6
7   def Pow(A,k):
8       if k == 1:
9           return(A)
10      if k % 2 != 0:
11          return(np.dot(A,(np.dot(Pow(A,(k-1)/2),Pow(A,(k-1)/2)))))
```

```
12     return(np.dot(Pow(A,k/2),Pow(A,k/2)))
13     if __name__ == '__main__':
14  B = np.array([[1,2,3],
15                [1,1,1],
16                [2,2,2]])
17  P = Pow(B,5)
18  print(P)
19  >>>
20  [[1041 1250 1459]
21   [ 521  625  729]
22   [1042 1250 1458]]
```

Finding nth power of a diagonal matrix is simply raising its diagonal elements to nth power. For example,

$$\begin{bmatrix} 3 & 0 & 0 \\ 0 & -1 & 0 \\ 0 & 0 & 2 \end{bmatrix}^3 = \begin{bmatrix} 3^3 & 0 & 0 \\ 0 & -1^3 & 0 \\ 0 & 0 & 2^3 \end{bmatrix} = \begin{bmatrix} 27 & 0 & 0 \\ 0 & -1 & 0 \\ 0 & 0 & 8 \end{bmatrix}$$

This property of a diagonal matrix may be used to find the power of a matrix. If $A = PDP^{-1}$ for some diagonal matrix D and some invertible matrix P, then,

$$A^2 = (PDP^{-1})(PDP^{-1}) = PD^2P^{-1}$$

and in general,

$$A^n = PD^nP^{-1}$$

The following properties of matrix powers may be stated.

- $A^r A^s = A^{r+s}$
- $(A^r)^s = A^{rs}$
- If A^n is invertible, then $A^{-n} = (A^{-1})^n$

3.10 Eigenvalues and Eigenvectors

The product of a matrix with a vector results in another vector as was noted. Let us consider the equation,

$$Ax = \lambda x \tag{3.1}$$

where A is a square matrix, x is a vector and λ is a real number. Thus, multiplication of vector x with matrix A has the same effect of multiplying x with a real number. When such an x vector is found, it is called the *eigenvector* of matrix A and λ is called the eigenvalue of A associated with that vector x. In general, there will be n eigenvectors and n corresponding eigenvalues of a square matrix of dimension n.

We will see how these parameters related to a matrix may be used for graph analysis later. Using Eq. 3.1; we can state the following,

$$A^2 x = \lambda A x = \lambda^2 x$$

In more general form,

$$A^n x = \lambda^n x$$

which means the eigenvectors of any power n of matrix A are the same and the eigenvalues of such a matrix are eigenvalues of A raised to nth power. Rewriting Eq. 3.1 yields,

$$(A - \lambda I)x = 0$$

where I is the identity vector of size n which means,

$$det(A - \lambda I) = 0$$

This polynomial equation of λ is called the *characteristic equation* of matrix A. Therefore, solving for the roots of this equation results in the eigenvalues of A after which eigenvectors may be computed.

Example 3.10.1 Let us find the eigenvalues and eigenvectors of the matrix

$$\begin{bmatrix} 0 & -1 \\ 2 & -3 \end{bmatrix}$$

The characteristic equation is,

$$(-\lambda)(-3 - \lambda) + 2 = \lambda^2 + 3\lambda + 2$$

which yields $\lambda_1 = -1$ and $\lambda_2 = -2$ as roots. Substituting these values in sequence in Eq. 3.1,

$$\begin{bmatrix} -\lambda & -1 \\ 2 & -3 - \lambda \end{bmatrix} \begin{bmatrix} x_1 \\ x_2 \end{bmatrix} = \begin{bmatrix} 0 \\ 0 \end{bmatrix}$$

and solving for x_1 and x_2 for $\lambda_1 = -1$ provides $x_1 = x_2$ which means there are infinite eigenvectors for this eigenvalue as below.

$$k \cdot \begin{bmatrix} 1 \\ 1 \end{bmatrix}$$

with $k \in \mathbb{R}$. For example, [3 3] is one such eigenvector. Substitution in Eq. 3.1 yields,

$$\begin{bmatrix} 0 & -1 \\ 2 & -3 \end{bmatrix} \begin{bmatrix} 3 \\ 3 \end{bmatrix} = -1 \cdot \begin{bmatrix} 3 \\ 3 \end{bmatrix}$$

Substituting $\lambda_2 = -2$ in Eq. 3.1 and solving for x_1 and x_2 of second eigenvector yields $2x_1 = x_2$. This time,

$$k \cdot \begin{bmatrix} 2 \\ 1 \end{bmatrix}$$

with $k \in \mathbb{R}$. Taking $x_1 = 2$ and $x_2 = 4$ yields,

$$\begin{bmatrix} 0 & -1 \\ 2 & -3 \end{bmatrix} \begin{bmatrix} 2 \\ 4 \end{bmatrix} = -2 \cdot \begin{bmatrix} 2 \\ 4 \end{bmatrix}$$

Some interesting properties of eigenvalues may be stated as follows.

- Determinant of a square matrix is the product of its eigenvalues.
- Trace of a square matrix is equal to the sum of its eigenvalues.

The method in Python to calculate the eigenvalues and eigenvectors of a matrix from the *numpy linalg* library is *eig*. It returns the eigenvalues and the eigenvectors as one eigenvector per column of a matrix as shown in the code below. The vector w contains the eigenvalues of the matrix A and the matrix V has one eigenvector per column. Eigenvalues for the given matrix A are 2, 1 and 4 and we can see that determinant of A is equal to the product of these values and the trace of A to the sum.

```
import numpy as np
from numpy import linalg as la
A = np.array([[2,2,-3],
              [0,2,-2],
              [0,-1,3]])
n = len(A)
vals, V = la.eig(A)
print("eigenvalues:",vals)
print("eigenvectors:\n",V)
print("Determinant of A: ", round(la.det(A)))
print("Trace of A: ", np.trace(A))
>>>
eigenvalues: [2. 1. 4.]
eigenvectors:
 [[ 1.          0.40824829  0.87038828]
 [ 0.         -0.81649658  0.34815531]
 [ 0.         -0.40824829 -0.34815531]]
Determinant of A:  8
Trace of A:  7
```

In order to extract the jth eigenvector from V, we can use

```
>>> u = V[:,j]
```

3.11   Block Matrices

A block matrix is defined in terms of smaller matrices. A block matrix M that consists of blocks A, B, C and D is shown below.

$$M = \begin{bmatrix} A & B \\ C & D \end{bmatrix}$$

Given the following matrices,

$$A = \begin{bmatrix} 1 & 3 \\ 0 & 1 \end{bmatrix}, B = \begin{bmatrix} 2 & 1 \\ 4 & 3 \end{bmatrix}, C = \begin{bmatrix} 1 & 0 & 2 \\ 2 & 3 & 1 \end{bmatrix}, D = \begin{bmatrix} 4 \\ 2 \end{bmatrix}$$

matrix M is as follows,

$$M = \begin{bmatrix} 1 & 3 & 2 & 1 \\ 0 & 1 & 4 & 3 \\ 1 & 0 & 2 & 4 \\ 2 & 3 & 1 & 2 \end{bmatrix}$$

Note that the blocks of a block matrix must fit together to form a rectangle. A matrix may be partitioned to its blocks using various methods. For example, if the matrix contains many zeros as in a sparse matrix, these parts may be partitioned into single blocks. We can perform various operations outlined for ordinary matrices by treating each block as a single matrix entry. For example, addition of two block matrices M_1 and M_2 may be performed by adding their related blocks as below.

$$\begin{bmatrix} A_1 & B_1 \\ C_1 & D_1 \end{bmatrix} + \begin{bmatrix} A_2 & B_2 \\ C_2 & D_2 \end{bmatrix} = \begin{bmatrix} A_1 + A_2 & B_1 + B_2 \\ C_1 + C_2 & D_1 + D_2 \end{bmatrix}$$

The product of these two matrices is given below.

$$\begin{bmatrix} A_1 A_2 + B_1 C_2 & A_1 B_2 + B_1 D_2 \\ C_1 A_2 + D_1 C_2 & C_1 B_2 + D_1 D_2 \end{bmatrix}$$

Thus, we multiply blocks of two block matrices as multiplying elements of two simple matrices, assuming blocks have compatible sizes.

3.12   Chapter Notes

We reviewed matrices, basic matrix operations, algorithms to realize these operations and implementations of these operations using Python in this chapter. The basic matrix operations are computing determinant, inverse of a matrix and matrix multiplication which is an important operation that finds a wide number of applications.

The Python library *numpy* provides most of the functions to realize these operations as shown. We will be using matrix multiplication in integer and boolean form when a graph is represented by a matrix and we need to implement a graph algorithm in chapters of Part II. A detailed analysis of matrix computations can be found in [1].

Exercises

1. Given the following matrix A, work out $A \cdot A^T$ and $A^T \cdot A$ and state any observable properties of these two products.

$$A = \begin{bmatrix} 1 & 2 & 3 \\ 4 & 5 & 6 \end{bmatrix}$$

2. Find the determinant of the matrix given below both by using calculation and Python *numpy* library.

$$\begin{bmatrix} 2 & 1 & 5 \\ 1 & 4 & 3 \\ 5 & 3 & 7 \end{bmatrix}$$

3. Write a Python function which inputs an integer matrix and returns it in upper triangular form and run this function for the following matrix.

$$\begin{bmatrix} 3 & 2 & 4 & 0 \\ 6 & 1 & 5 & 2 \\ 5 & 3 & 7 & 8 \\ 1 & 6 & 4 & 2 \end{bmatrix}$$

4. Write a Python function that inputs two compatible matrices, computes their product and returns this product.
5. Find the product of he two matrices given below using the *numpy* method *dot*.

$$\begin{bmatrix} 3 & 1 & 4 \\ 2 & 0 & 5 \end{bmatrix} \begin{bmatrix} 2 & 1 \\ 1 & 3 \\ 7 & 2 \end{bmatrix}$$

6. Find the eigenvalues and the eigenvectors of the following matrix by calculation and also using the *numpy* method *eig*.

$$\begin{bmatrix} 0 & 1 \\ -2 & 3 \end{bmatrix}$$

Reference

1. G.H. Golub, G.F. Van Loan, *Matrix Computations* (Johns Hopkins Studies in the Mathematical Sciences), 4th edn. (2013)

Graphs, Matrices and Matroids

4

Abstract

This chapter forms the basic background on graphs, matrices and matroids. We start with the definitions and notations of graphs followed by main graph types, graph operations and graph traversals in the first section. We then look at ways of defining graph in terms of matrices and conclude with a data structure called matroids which we will use to design some specific type of graph algorithms.

4.1 Introduction to Graphs

A graph denoted by $G = (V, E)$ consists of a set V of vertices and a set E of edges between the vertices. A graph is simple when the number of edges between any of its vertices is at most 1 and it has no self-loops around any of its vertices. We will consider mostly simple graphs in this text. A simple graph $G = (V, E)$ is shown in Fig. 4.1 with $V = \{a, b, c, d, e, f\}$ and $E = \{e_1, e_2, e_3, e_4, e_5, e_6, e_7, e_8, e_9\}$. We will show an edge as (u, v) when such an edge is incident on the vertices u and v, for example, $e_1 = (f, a)$ in this figure.

The vertices of a graph are commonly named *nodes* or *endpoints* of an edge when an edge is referred. The number of vertices of a graph is called its *order* and the number of edges it has is referred to as its *size*. We will show the order of a graph with the literal n and its size with m as in common use. We will consider finite graphs in this text which have a finite number of vertices and edges.

The complement of a graph $G = (V, E)$, $\overline{G} = (V, E')$, has the same vertex set as G but lack of an edge in G is represented by an edge in $\overline{G}$. Formally, if $(u, v) \in E$, then $(u, v) \notin E'$ and vice versa. A sample graph and its complement are depicted in Fig. 4.2.

© Springer Nature Switzerland AG 2021
K. Erciyes, *Algebraic Graph Algorithms*, Undergraduate Topics in Computer Science,
https://doi.org/10.1007/978-3-030-87886-3_4

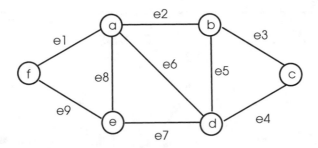

Fig. 4.1 An example undirected graph

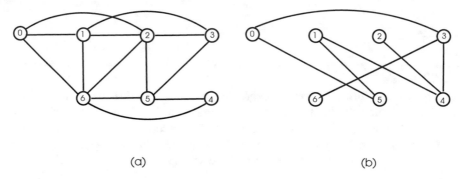

(a)   (b)

Fig. 4.2 **a** An example graph, **b** Its complement graph

4.1.1 Degree of a Vertex

Definition 4.1 (*degree of a vertex*) The degree of a vertex v in a graph G is the sum of the number of proper edges and twice the number of self-loops incident on v. This parameter is commonly denoted by $deg(v)$ for vertex v.

The maximum degree in a graph G is shown by $\Delta(G)$ and the minimum degree is denoted by $\delta(G)$. These parameters are 4 and 2 respectively for the graph of Fig. 4.1. A vertex of degree of 0 in a graph is called an *isolated* vertex and a vertex of degree 1 is called a *pendant* vertex. Based on foregoing, the following can be stated for any vertex v of a graph G:

$$0 \le \delta(G) \le deg(v) \le \Delta(G) \le n - 1 \tag{4.1}$$

The sum of the degrees of a graph is an even number since we count each edge twice, once from each end, when calculating this sum. The sum of the degrees of vertices in the graph of Fig. 4.1 is 18. We can now state the following equation to show the relation between the sum of degrees of a graph and the its number of edges m.

$$\sum_{v \in V} deg(v) = 2m$$

The average degree of a graph, $deg(G)$ is then $2m/n$. The neighborhood $N(v)$ of a vertex v in a graph is the set of vertices that are adjacent to v. For example, $N(a) = \{b, d, e, f\}$ in the graph of Fig. 4.1.

4.1.1.1 Degree Sequences

The degree sequence of a graph G is a list containing degrees of vertices of G in non-decreasing or non-increasing order. The degree sequence of the graph of Fig. 4.1 is $\{2, 2, 3, 3, 4, 4\}$ in non-decreasing order. If a given a list D of integers represents a real graph, D is called *graphical*. Note that a simple way to test whether D is graphical is checking whether the sum is an even number. For example $D = \{1, 3, 4, 5\}$ is not graphical meaning we can not realize this degree sequence in a graph. It is possible to have different graphs with the same degree sequence.

4.1.2 Subgraphs

Given a graph $G = (V, E)$, a subgraph of G contains a subset of its vertices and/or a subset of its edges. Formally, if $G' = (V', E')$ is a subgraph of G, then $V' \subseteq V$ and/or $E' \subseteq E$. Whenever G' is not equal to G, G' is called a *proper subgraph* of G. If G' contains all edges of G, then G is called an *induced subgraph* of G. On the other hand, if G' contains all vertices of G, G' is called a *spanning subgraph* of G. When viewed from G, G is the *supergraph* of G' and called the *spanning supergraph* of G' when they both have the same vertex set.

An induced subgraph of G that consists of vertices in set V' is shown by $G[V']$ and the subgraph obtained by deleting vertices in the set V' is denoted by $G - V'$. A sample graph is depicted in Fig. 4.3a, its spanning subgraph is shown in (b) and an induced subgraph of this graph is given in (c).

4.2 Graph Operations

A new graph may be generated by performing some operation on two or more graphs. Some common operations of graphs are the union, intersection and cartesian product. The *union* of two graphs $G_1 = (V_1, E_1)$ and $G_2 = (V_2, E_2)$ is a graph $G_3 = (V_3, E_3)$ in which $V_3 = V_1 \cup V_2$ and $E_3 = E_1 \cup E_2$. Informally, the union of $G1$ and $G2$ contains the union of the two sets of vertices and two sets of edges in these two graphs. Figure 4.4c depicts the union of two graphs $G1$ and $G2$ in (a) and (b).

The *intersection* of two graphs $G_1 = (V_1, E_1)$ and $G_2 = (V_2, E_2)$ is a graph $G_3 = (V_3, E_3)$ in which $V_3 = V_1 \cap V_2$ and $E_3 = E_1 \cap E_2$. In other words, the

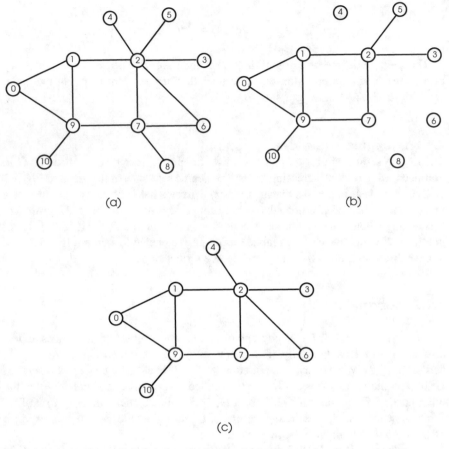

Fig. 4.3 a A sample graph G, **b** A spanning subgraph of G, **c** An induced subgraph of G

union of the intersection of $G1$ and $G2$ contains the intersection of the vertices and intersection of edges in these two graphs. The intersection of two graphs is shown in Fig. 4.4d.

The cartesian product of two graphs $G_1 = (V_1, E_1)$ and $G_2 = (V_2, E_2)$ denoted by $G_1 \square G_2$ or $G_1 \times G_2$ is a graph $G_3 = (V_3, E_3)$ where $V_3 = V_1 \times V_2$ and an edge $((u_i, v_j), (u_k, j_l))$ is in $G_1 \times G_2$ if one of the following conditions holds:

- $i = k$ and $(v_j, v_l) \in E_2$
- $j = l$ and $(u_i, u_k) \in E_1$

We have a vertex representing two vertices, one from each graph, in the cartesian product of two graphs. The cartesian product of two linear graphs of three and two vertices of Fig. 4.5a is depicted in (b) of the same figure with solid edges representing the first condition above and the dashed edges are formed using the second condition.

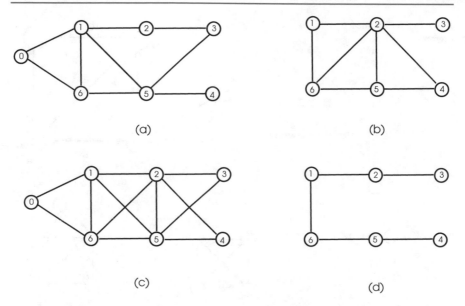

Fig. 4.4   **a** A sample graph $G1$, **b** Another sample graph $G2$, **c** Union of $G1$ and $G2$, **d** Intersection of $G1$ and $G2$

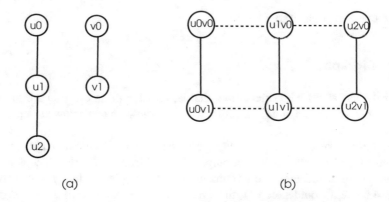

Fig. 4.5   **a** Linear graphs with three and two vertices, **b** Cartesian product of these graphs

4.3   Types of Graphs

Some types of graphs are more frequently implemented to model real networks. We review several important types of graphs in this section.

Fig. 4.6 An example
digraph

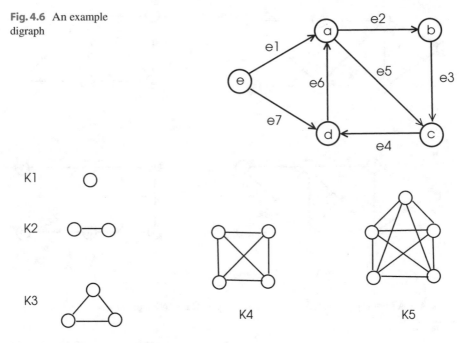

Fig. 4.7 Complete graphs of sizes 1 to 5

4.3.1 Digraph

A *digraph* has edges with orientation, that is, an edge or an *arc e* in a digraph *G*
starts from a vertex and ends in a vertex which is shown by an arrow as depicted in
Fig. 4.6.

The *indegree* of a vertex v in a digraph is the number of ingoing edges to v and
the outdegree of v is the number of outgoing edges from that vertex. The degree of
v is the sum of its indegree and its outdegree. For example, the indegree of vertex c
in Fig. 4.6 is 2, its outdegree is 1, thus, its degree is 3. A *complete simple digraph*
has a pair of oriented edges, one for each direction, between all vertex pairs.

4.3.2 Complete Graphs

A *complete simple graph* has edges between every vertex pairs. A complete graph
that has n vertices is depicted by K_n, for example, K_4 is a complete graph with four
vertices. The size of a simple undirected complete graph K_n is $n(n-1)/2$ since
there are n vertices with $n-1$ degree each and each edge (u, v) is counted twice
one for vertex u and one for vertex v. The complete graphs K_1, K_2, K_3, K_4 and K_5
are shown in Fig. 4.7.

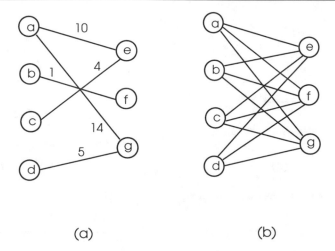

Fig. 4.8 **a** A weighted undirected bipartite graph, **b** A complete bipartite graph with the same vertex set

4.3.3 Weighted Graphs

Edges have weights associated with edges in an edge-weighted graph $G(V, E, w)$, $w : E \rightarrow \mathbb{R}$ and a vertex-weighted graph $G(V, E, w)$, $w : V \rightarrow \mathbb{R}$ has vertices with labels representing weights.

There are many real-life networks represented by weighted graphs, for example, a road network may be modeled conveniently by a weighted graph where distance between two cities is represented by the weight of an edge, and a computer network with link capacities may be represented by a weighted graph. The vertex weights in a vertex weighted graph may show the size of a protein in a protein interaction network.

4.3.4 Bipartite Graphs

A bipartite graph G has two distinct vertex sets V_1 and V_2 such that any edge in G is has one endpoint in $V1$ and the other endpoint in V_2. A bipartite graph may be weighted with edges and/or vertices having weights; it may be directed or it may be complete with edges between every vertex pairs. A complete bipartite graph is denoted by $K_{p,q}$ where p is the number of vertices in the first vertex set and q holds the number of vertices in the second. A bipartite graph may be used to model various problems such as scheduling of tasks to processor where vertices in V_1 represents the tasks and vertices in V_2 are the processors. An undirected weighted bipartite graph is depicted in Fig. 4.8a and a complete unweighted bipartite graph $K_{4,3}$ with the same vertex set is shown in (b).

4.3.5 Regular Graphs

A *k-regular* graph has vertices all with a degree of k. Regular graphs with degrees 1, 2, and 3 are shown in Fig. 4.9. The 3-regular graph shown in (d) is known as the

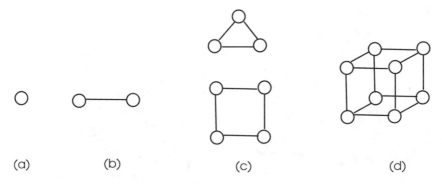

Fig. 4.9 **a** 0-regular, **b** 1-regular, **c** 2-regular graphs; **d** A 3-regular graph

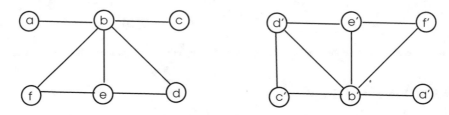

Fig. 4.10 Two isomorphic graphs where a vertex x is mapped to vertex x'

hypercube and is used as an architecture for parallel processing where the vertices are the computing elements and the edges are the communication links. An n-cycle graph is a 2-regular graph as the graphs of (b) in this figure.

4.3.6 Graph Isomorphism

Two graphs G_1 and G_2 are isomorphic if they have the same number of vertices and edges, and their edge connectivity is preserved. Formally, an isomorphism from a graph G_1 to another graph G_2 is a function $f : V(G_1) \rightarrow V(G_2)$ where an edge $(u, v) \in E(G_1)$ if and only if $f(u)f(v) \in E(G_2)$.

Two isomorphic graphs are depicted in Fig. 4.10. Testing whether two graphs are isomorphic is a difficult problem and cannot be performed in polynomial time. An isomorphism of a graph to itself is called an *automorphism*.

4.4 Walks, Paths, Cycles

Graph traversals may be explicitly defined using following terminology.

- *Walk*: A walk W between two vertices v_i and v_j of a graph G is an alternating sequence of vertices and edges between these two vertices. The vertex v_i is called the *initial* vertex and v_j is called the *terminating* vertex of the walk W. A walk can contain repeated vertices and edges. The length of a walk is the number of edges contained in it. A directed walk in a digraph obeys the direction of edges.
- *Trail*: A trail in a graph is a walk that does not have any repeated edges.
- *Path*: A path of a graph is a trail that does not include any repeated vertices other than the initial and terminal vertices. A path is shown by the sequence of vertices it contains including the starting and ending vertices. The length of a path is the number of vertices contained in it.
- *Cycle*: A cycle is a path which starts and ends at the same vertex. A cycle is called an *odd-cycle* if the length of a cycle is an odd number, it is called an *even-cycle* otherwise.
- *Circuit*: A circuit is a trail that starts and ends at the same vertex.
- *Connectivity*: A graph is called *connected* if there is a walk between any vertex pairs it contains.

A *Hamiltonian Cycle* in a graph goes once through each vertex of the graph and a graph with such property is called *Hamiltonian*. An *Eulerian cycle* is a cycle of a graph that goes exactly once through each edge in the graph. We will use the graph of Fig. 4.11 to illustrate these concepts as below.

- $\{a, e_0, b, e_2, i, e_1 1\}$ is a walk.
- $\{b, e_1, c, e_3, g, e_2, b, e_0, a\}$ is a trail that starts at vertex b and ends at vertex a.
- (b, c, f, e) is a path that starts at vertex b and ends at vertex e.
- $\{c, d, e, f, g, c\}$ is a cycle.
- $\{a, g, f, e, d, c, f, b, c, g, b, a\}$ is a Hamiltonian cycle. Note that vertices a, b, c and f are visited twice but each edge of the graph is visited exactly once in this cycle.

The length of the shortest path between two vertices u and v of a graph is termed as the distance between the vertices u and v. For example, $a, b, f, c, a, b, f, c,$ a, g, f, e, d, c and a, g, c are all paths between vertices a and c in the graph of

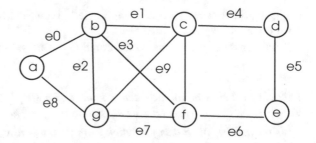

Fig. 4.11 A sample graph

Fig. 4.11 with lengths 3, 3, 5 and 2 respectively, thus, distance between these two vertices is the length of the smallest length path between them which is 2 in this case.

The maximum distance of a vertex v to any other vertex in a graph is called the *eccentricity* of v and the maximum eccentricity of vertices in a graph G is called the *diameter* of the graph. This parameter shows the easiness of reaching any two farthest points in the graph, for example, a low diameter network allows fast communication between any of its nodes. The radius of a graph is the minimum value of the eccentricities of its vertices. The *center* of a graph is the vertex with minimum eccentricity, clearly, there may be more than one center in a graph. Finding centers in a graph that represents some real-life network is beneficial as it helps to locate shared resources in places to be easily reached by consumers.

4.5 Graphs and Matrices

In this section, we review some important matrices that are used to represent graphs with properties that enable to extract information about the structures of graphs. Graphs are mainly represented using the *adjacency matrix*, the *incidence matrix* and the *adjacency list* methods. There are various other matrices associated with graphs as we will see.

4.5.1 Adjacency Matrix

An *adjacency matrix* of a simple graph or a digraph is a matrix $A_{n \times n}$ where entry a_{ij} of this matrix equals 1 if node i is connected to node j, otherwise a_{ij} equals 0. The entry a_{ij} of this matrix when representing a digraph is 1 if there is an outgoing edge from vertex i to vertex j. For multigraphs, a_{ij} equals the number of edges from vertex i to vertex j. Clearly, this way of representing a graph requires $O(n^2)$ space. However, checking the existence of an edge (i, j) requires $O(1)$ operation by simply testing the value of a_{ij}. The adjacency matrix of the digraph of Fig. 4.6 is depicted in Fig. 4.12. Note that a_{ij} is 1 when the edge between i and j starts from node i and ends at node j. The adjacency matrix A has the following properties.

- The matrix A is symmetric for undirected graphs and is not symmetric in general for directed graphs
- A has zeros on its diagonal when graph G has no self-loops.

Theorem 1 *Let A be the adjacency matrix of a simple graph G with a vertex set $V = \{v_1, v_2, \ldots, v_n\}$ and let k be a positive integer. Then the entry a_{ij}^k of the matrix A^k is the number of walks of length k from vertex v_i to v_k in G.*

Fig. 4.12 Adjacency matrix
of the graph of Fig. 4.6

$$
\begin{array}{c c}
 & \begin{array}{c c c c c} a & b & c & d & e \end{array} \\
\begin{array}{c} a \\ b \\ c \\ d \\ e \end{array} &
\left(
\begin{array}{c c c c c}
0 & 1 & 1 & 0 & 0 \\
0 & 0 & 1 & 0 & 0 \\
0 & 0 & 0 & 1 & 0 \\
1 & 0 & 0 & 0 & 0 \\
1 & 0 & 0 & 1 & 0
\end{array}
\right)
\end{array}
$$

Fig. 4.13 Adjacency list of
the graph of Fig. 4.6

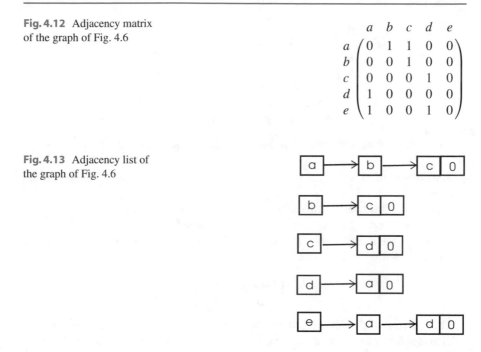

4.5.2   Adjacency List

An *adjacency list* of a simple graph or a digraph consists of multiple linked lists of
its nodes. The neighbors of a node *a* are included in its linked list with each node
pointing to another neighbor of *a* and the end node of the list points to *null* to show
it is not pointing any other node. Commonly, the addresses of starting nodes are kept
in an array of pointers. The adjacency list of the digraph of Fig. 4.6 is shown in
Fig. 4.13. Finding an element of a graph in this notation requires $O(m)$ time and the
space needed is $O(n + m)$. We will see that storing of a sparse graph is commonly
performed by some modified adjacency list representation.

4.5.3   Incidence Matrix

An *incident matrix* $B(G)$ of a simple undirected graph has elements $b_{ij} = 1$ if edge j
is incident to vertex i and $b_{ij} = 0$ otherwise. We need to consider the starting vertex
and the ending vertex of an edge when defining the incidence matrix of a digraph as
below.

$$
b_{ve} = \begin{cases}
1 & \text{if edge } e \text{ starts at vertex } v \\
-1 & \text{if edge } e \text{ ends at vertex } v \\
0 & \text{otherwise}
\end{cases}
$$

The incidence matrix of the graph of Fig. 4.6 is below.

$$
\begin{array}{c c c c c c c c}
 & e_1 & e_2 & e_3 & e_4 & e_5 & e_6 & e_7 \\
a & -1 & 1 & 0 & 0 & 1 & -1 & 0 \\
b & 0 & -1 & 1 & 0 & 0 & 0 & 0 \\
c & 0 & 0 & -1 & 1 & -1 & 0 & 0 \\
d & 0 & 0 & 0 & -1 & 0 & 1 & -1 \\
e & 1 & 0 & 0 & 0 & 0 & 0 & 1
\end{array}
$$

The following are observed in an incidence matrix of any simple graph.

- The number of 1s in a row i is equal to the degree (outdegree in a digraph) of the vertex i.
- A column j has two 1s since edge j is incident on exactly two vertices.
- An isolated vertex i is manifested by all zeros in row i.
- The incidence matrix $B(G)$ of a graph G with two components G_1 and G_2 can be specified as follows.

$$
\begin{bmatrix}
B(G_1) & 0 \\
0 & B(G_2)
\end{bmatrix}
$$

where $B(G_1)$ and $B(G_2)$ are the incidence matrices of components G_1 and G_2. This is valid since edges of G_1 are not incident to any vertex of G_2 and vice versa.

4.5.4 Cycle Matrix

Definition 4.2 (*cycle matrix*) The cycle matrix B of a graph G is a matrix of cycles of G as its rows and the edges of G as its columns. The entry b_{ij} of $B(G)$ is defined as follows.

$$
b_{ij} = \begin{cases} 1 & \text{if } i\text{th cycle is incident on edge } j \\ 0 & \text{otherwise} \end{cases}
$$

Let us consider the graph of Fig. 4.14. The following cycles are present in this graph,

- $c_1 = \{e_1, e_6, e_7\}$
- $c_2 = \{e_1, e_2, e_5, e_7\}$
- $c_3 = \{e_2, e_5, e_6\}$
- $c_4 = \{e_3, e_4, e_5\}$
- $c_5 = \{e_1, e_3, e_4, e_6\}$

The cycle matrix of this graph, is represented in Fig. 4.15 with cycles as rows and edges as columns. The following properties of the cycle matrix $C(G)$ of a graph G can be observed.

- All 0s in the jth column of C means edge j does not lie on any cycle of G.

Fig. 4.14 An example graph
G

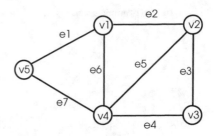

Fig. 4.15 Its cycle matrix

$$
\begin{array}{c}
\\
c_1 \\
c_2 \\
c_3 \\
c_4 \\
c_5
\end{array}
\begin{array}{ccccccc}
e_1 & e_2 & e_3 & e_4 & e_5 & e_6 & e_7 \\
\left(\begin{array}{ccccccc}
1 & 0 & 0 & 0 & 0 & 1 & 1 \\
1 & 1 & 0 & 0 & 1 & 0 & 1 \\
0 & 1 & 0 & 0 & 1 & 0 & 0 \\
0 & 0 & 1 & 1 & 1 & 0 & 0 \\
1 & 0 & 1 & 1 & 0 & 0 & 1
\end{array}\right)
\end{array}
$$

- The number of 1s in row i of G is equal to the number of edges in cycle i of G.
- A self-loop on a vertex j of G is manifested by a single 1 in the column j of $C(G)$.
- Permutation of any rows or columns of C corresponds to re-labeling of the cycles or edges in $C(G)$.

4.5.5 Path Matrix

Definition 4.3 (*path matrix*) Let G be a graph with m edges and u and v be any vertex pair in G. The *path matrix* of vertices u and v, $P(u, v) = [p_{ij}]$ specifies the number of different paths between u and v as follows,

$$
p_{ij} = \begin{cases} 1 \text{ if } j\text{th edge is incident on path } i \\ 0 \qquad\qquad \text{otherwise} \end{cases}
$$

A path matrix provides the paths between a selected pair of vertices. Considering the graph of Fig. 4.16, the following paths between the vertices v_6 and v_3 exist.

- $p_1 = \{e_7, e_6, e_2\}$
- $p_2 = \{e_7, e_6, e_5, e_3\}$
- $p_3 = \{e_7, e_4, e_3\}$
- $p_4 = \{e_7, e_4, e_5, e_2\}$

We can then fill the entries of the path matrix $P(v_6, v_3)$ as shown in Fig. 4.17. The following properties of the cycle matrix $C(G)$ of a graph G can be observed.

- All 0s in the jth column of C means edge j does not lie on any cycle of G.

Fig. 4.16 An example graph *G*

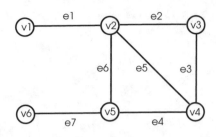

Fig. 4.17 Path matrix $P(v_6, v_3)$

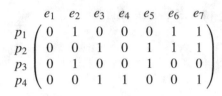

Fig. 4.18 A sample graph

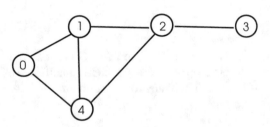

- The number of 1s in row i of G is equal to the number of edges in cycle i of G.
- A self-loop on a vertex j of G is manifested by a single 1 in the column j of $C(G)$.
- Permutation of any rows or columns of C corresponds to re-labeling of the cycles or edges in $C(G)$.

4.5.6 The Laplacian

The *Laplacian matrix* L of a graph G is $L = D(G) - A(G)$ where $D(G)$ is a diagonal matrix of G with vertex degrees in its diagonal such that d_{ii} is the degree of vertex i. The element l_{ij} of this matrix can be formed as follows.

$$l_{ij} = \begin{cases} -1 & \text{if } i \neq j \text{ and } a_{ij} = 1 \\ 0 & \text{if } i \neq j \text{ and } a_{ij} = 0 \\ -d_{ij} & \text{if } i = j \end{cases}$$

where d_{ii} is the diagonal element of matrix D. The adjacency matrix A and the diagonal matrix D of the graph of Fig. 4.18 is shown below.

$$A = \begin{array}{c} \\ 0 \\ 1 \\ 2 \\ 3 \\ 4 \end{array} \begin{array}{ccccc} 0 & 1 & 2 & 3 & 4 \\ \begin{pmatrix} 0 & 1 & 0 & 0 & 1 \\ 1 & 0 & 1 & 0 & 1 \\ 0 & 1 & 0 & 1 & 1 \\ 0 & 0 & 1 & 0 & 0 \\ 1 & 1 & 1 & 0 & 0 \end{pmatrix} \end{array}, D = \begin{array}{c} \\ 0 \\ 1 \\ 2 \\ 3 \\ 4 \end{array} \begin{array}{ccccc} 0 & 1 & 2 & 3 & 4 \\ \begin{pmatrix} 2 & 0 & 0 & 0 & 0 \\ 0 & 3 & 0 & 0 & 0 \\ 0 & 0 & 3 & 0 & 0 \\ 0 & 0 & 0 & 1 & 0 \\ 0 & 0 & 0 & 0 & 3 \end{pmatrix} \end{array}$$

The Laplacian matrix for this graph is calculated as below.

$$L = \begin{array}{c} \\ 0 \\ 1 \\ 2 \\ 3 \\ 4 \end{array} \begin{array}{ccccc} 0 & 1 & 2 & 3 & 4 \\ \begin{pmatrix} 2 & -1 & 0 & 0 & -1 \\ -1 & 3 & -1 & 0 & -1 \\ 0 & -1 & 3 & -1 & -1 \\ 0 & 0 & -1 & 1 & 0 \\ -1 & -1 & -1 & 0 & 3 \end{pmatrix} \end{array}$$

The normalized *Laplacian matrix* $\mathcal{L}$ is given by the below equation.

$$\mathcal{L} = D^{-1/2} L D^{-1/2} = D^{-1/2}(D - A)D^{-1/2} = I - D^{-1/2} A D^{-1/2} \qquad (4.2)$$

The normalized Laplacian then has the following elements.

$$\mathcal{L}_{ij} = \begin{cases} 1 & \text{if} & i = j \\ \frac{-1}{\sqrt{d_i d_j}} & \text{if} & i \text{ and } j \text{ are neighbors} \\ 0 & \text{otherwise} \end{cases}$$

4.6 Python Graphs

Python provides the *draw* method from the *networkx* library to draw a graph. In one of the simple ways to generate a graph, we first define a graph object using the *Graph* method from this library and then can add edges to this object using the *add_edges_from* method from a list which specifies edges as vertex pairs as in the example below. Plotting of the drawn graph can be performed by *show* method from the *matplot.pyplot* library.

```
import networkx as nx
import matplotlib.pyplot as plt

G= nx.Graph()
L = [(0,1),(1,2),(0,6),(5,6),(1,6),(2,5),(2,3),(2,4),(2,6),(6,7)]
G.add_edges_from(L)
nx.draw(G,with_labels=1)
plt.show()
```

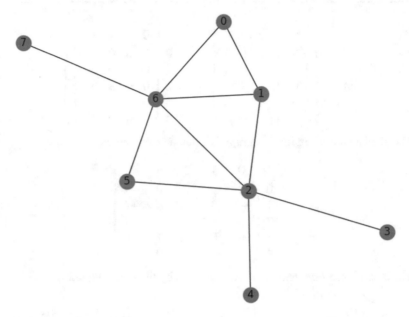

Fig. 4.19 A sample graph generated by Python *pyplot*

The plotted graph when this program is run is depicted in Fig. 4.19. A directed graph can be drawn by the *digraph* method from the *networkx* library as in the following code.

```
import networkx as nx
import matplotlib.pyplot as plt

G= nx.DiGraph()
L = [(0,1), (1,2),(2,4),(3,4),(4,5),(5,2),(4,1)]
G.add_edges_from(L)
nx.draw(G,with_labels=1)
plt.show()
```

The digraph plotted using the edge list in this code is shown in Fig. 4.20. It is possible to convert an edge list of a graph to its adjacency matrix by simply by iterating through the edge list for each edge (i, j) and setting $a_{ij} = a_{ji} = 1$ in the matrix A for an undirected graph and setting just $a_{ij} = 1$ for a directed graph. The reverse operation of obtaining an edge list from the adjacency matrix is also possible by checking each entry a_{ij} of A and whenever this entry is 1, and edge (i, j) may be appended to the edge list. Note that the upper triangle of A is sufficient for an undirected graph but the whole of matrix A needs to be investigated for a directed graph. The numpy library provides the method $from\_numpy\_matrix$ to create a graph object directly from its adjacency matrix which can then be plotted as in the following example. The plotted graph is depicted in Fig. 4.21.

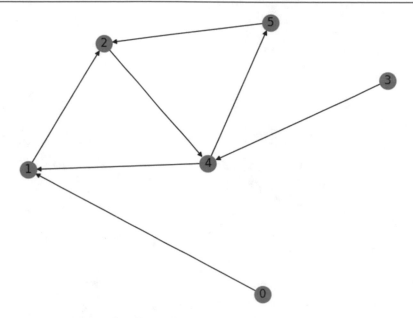

Fig. 4.20 A sample digraph generated by Python

```
import numpy as np
import networkx as nx
import matplotlib.pyplot as plt

A = np.array([[0,1,1,0,0,0],
              [1,0,1,1,0,0],
              [1,1,0,0,0,0],
              [0,1,0,0,1,1],
              [0,0,0,1,0,1],
              [0,0,0,1,1,0]])
G = nx.from_numpy_matrix(A)
nx.draw(G,with_labels=1)
plt.show()
```

4.7 Matroids

Matroid theory is a powerful paradigm that can be applied to solve various graph problems. A matroid description of a graph problem commonly ends in a greedy algorithm to solve that problem. The main forms of a matroid are the basic matroid, the graphic matroid and the weighted matroid.

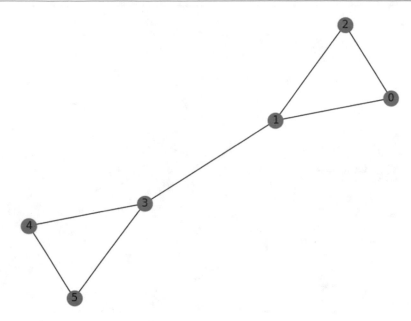

Fig. 4.21 A sample graph generated from its adjacency matrix

4.7.1 The Basic Matroid

Let us briefly review matroids;

Definition 4.4 (*matroid*) $M = (E, \mathcal{I})$ is a matroid if the following two conditions hold.

1. $\mathcal{I} \leftarrow \emptyset$
2. *Hereditary property*: If $B \in \mathcal{I}$ and $A \subset B$, then $A \in \mathcal{I}$. The set $\mathcal{I}$ is called hereditary.
3. *Exchange property*: If $A, B \in \mathcal{I}$ and $|B| > |A|$, then $\exists b \in \{B \setminus A\}$ such that $A \cup \{b\} \in \mathcal{I}$.

The subsets of $\mathcal{I}$ are called *independent sets*, and the maximal independent sets are called the *bases* of $\mathcal{M}$. The *rank* of an arbitrary subset $S \subseteq E$ is the size of the largest independent set contained in E.

Example 4.7.1 Let $E = \{1, 2, 3\}$ and $\mathcal{I} = \{\{1, 2\}, \{2, 3\}, \{1\}, \{2\}, \{3\}, \emptyset\}$. Then $\mathcal{M} = (E, \mathcal{I})$ is a matroid.

4.7.2 The Graphic Matroid

Definition 4.5 Let $G = (V, E)$ be a non-empty, undirected simple graph and $M_G = (S_G, \mathcal{I}$ be the matroid associated with G such that,

$$S_G = E$$

$$\mathcal{I} = \{A : A \subseteq E, (V, A) \text{ is a forest}\}$$

Such a matroid is termed the graphic matroid.

The independent sets in the graphic matroid are sets of edges of spanning forests of G. The following theorem provides the basic link between the matroids and graphs and can be used to design various graph algorithms as we will see. It basically shows that forests of a graph form a matroid.

Theorem 4.1 *For a graph $G = (V, E)$, a forest F is any set of edges of G that does not contain any cycles. $M = (E, \mathcal{F})$ where $\mathcal{F} = \{F \subseteq E : F \text{ is a forest of } G\}$ is a matroid.*

Proof Let us check the properties of being a matroid.

1. F is a nonempty hereditary system.
2. Let $A \subset F$ and $B \subset F$ with $|A| < |B|$. Then the forest $F_1 = (V, B)$ has less trees than the forest $F_2 = (V, A)$. Therefore, F_1 contains some trees that are in different trees in F_1. We can therefore add an edge e that connects two such trees to obtain another forest $F_3 = (V, A \cup \{e\})$. □

4.7.3 Weighted Matroid

Let us first define a weighted matroid to help us matroid-based design of tree algorithms.

Definition 4.6 (*weighted matroid*) A matroid $\mathcal{M} = (E, F)$ is called *weighted* if it has a weight function $w : E \to R^+$ associated with it.

A greedy algorithm to find the maximal independent sets for matroids can be implemented with following steps.

1. Let $\mathcal{M} = (E, F)$ be a matroid with a weight function w.
2. $A \leftarrow \varnothing$
3. Sort E in decreasing order by weight w.
4. **For all** $u \in E$ taken in decreasing order
5.   **If** $A \cup \{u\} \in F$ **then** $A \leftarrow A \cup \{u\}$
6. **End For**
7. **return** A

Fig. 4.22 An example graph

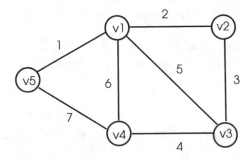

The main conclusion to be drawn is that the greedy algorithms for matroids do produce optimal results.

4.7.4   Graphs, Matroids and Matrices

We will describe the relationship between a graph and a matroid by a simple example. Considering the graph $G = (V, E)$ in Fig. 4.22, the set $\mathcal{C}$ of edge-sets of cycles in G is as follows.

$$\mathcal{C} = \{\{1, , 6, 7\}, \{2, 3, 5\}, \{4, 5, 6\}, \{1, 5, 4, 7\}, \{2, 3, 4, 6\}, \{1, 2, 3, 4, 7\}\}$$

Then, $M = (E, \mathcal{C})$ is called the *cycle matroid* of G and is denoted by $M(G)$. We can now have an alternative definition of a graphic matroid as below.

Definition 4.7 (*graphic matroid*) A matroid M with a ground set $E(M)$ is graphic if it is isomorphic to the cycle matroid of some graph.

4.8   Chapter Notes

This chapter serves as the basic background on graphs, graph matrices and matroids. We reviewed fundamental graph theory concepts in the first part with focus on types of graphs that are used frequently and graph traversals. A thorough treatment of graph theory can be found in [1–5]. We then looked at ways of matrix representations of graphs which are the adjacency, incidence, cycle, path and Laplacian matrices. We will be implementing most of the algebraic graph algorithms using the adjacency matrix and in few cases the incidence matrix. The Laplacian matrix of a graph provides information on the connectivity of a graph and also used to partition a graph. We then illustrated ways of representing and drawing graphs in the Python language. In the last part of the chapter, we reviewed a structure called a matroid which can be used to solve various graph problems such as implementing greedy graph algorithms.

Exercises

1. Work out the complement of the graph of Fig. 4.23.
2. Find the degree sequence and the degree matrix of the graph of Fig. 4.23.
3. Find the union and the intersection of the two graphs shown in Fig. 4.24.
4. Work out the cartesian product of the two graphs shown in Fig. 4.24.
5. Determine the number of edges of a k-regular graph that has n vertices.
6. Prove that a regular bipartite graph $G(V1, V2, E)$ has $|V1| = |V2|$.
7. Let A, B and C be the vertex sets with identifiers $\{0, 1, 2\}$, $\{3, 4, 5, 6\}$ and $\{7, 8\}$ respectively. Let G_1 be the complete A, B-bipartite graph and G_2 be the complete B, C-bipartite graph. Draw the figures of the graphs $G_1 \cup G_2$ and $G_1 \cap G_2$.

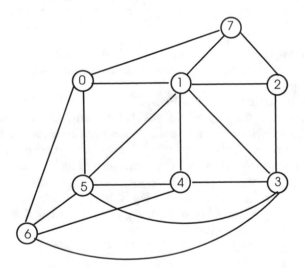

Fig. 4.23 A sample graph for Exercise 1

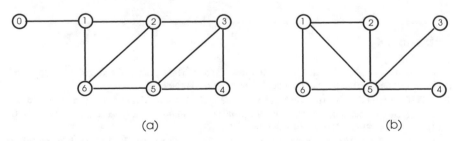

(a)   (b)

Fig. 4.24 A sample graph for Exercise 3

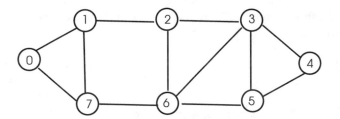

Fig. 4.25 A sample graph for Exercises 9 and 10

Fig. 4.26 A sample graph
for Exercise 11

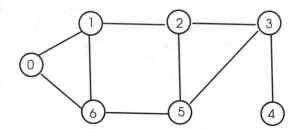

8. Draw the complete bipartite graph $K_{4,5}$ and work out the number of edges in this graph. Generalize this result to determine the number of edges in a complete bipartite graph.
9. Find all the cycles in the graph of Fig. 4.25.
10. Determine whether the graph of Fig. 4.25 is Eulerian and/or Hamiltonian.
11. Find the adjacency, incidence and the Laplacian matrices of the graph of Fig. 4.26.
12. A distance or weighted adjacency matrix of a weighted graph has a similar structure to its adjacency matrix with a 1 between vertex i and vertex j is replaced by the weight of the edge (i, j). Work out the adjacency matrix, incidence matrix and the distance matrix of the graph of Fig. 4.8a.
13. Write a Python program to draw K_4 using edge list. Also, draw K_4 from the adjacency matrix using Python.
14. Let $E = \{1, 2, 3\}$ and $\mathcal{I} = \{\{1, 2\}, \{2, 3\}, \{1\}, \{2\}, \{3\}, \{4\}, \varnothing\}$. Show that $\mathcal{M} = (E, \mathcal{I})$ is not a matroid.

References

1. B. Bollobas, *Modern Graph Theory*. Graduate Texts in Mathematics, Corrected edn. (Springer, Berlin, 2002). ISBN-10: 0387984887 ISBN-13: 978-0387984889
2. A.B. Bondy, U.S.R. Murty, *Graph Theory*. Graduate Texts in Mathematics (Springer, Berlin, 2008); 1st Corrected ed. 2008. Corr. 3rd printing 2008 edition (August 28, 2008). ISBN-10: 1846289696 ISBN-13: 978-1846289699
3. R. Diestel, *Graph Theory*. Graduate Texts in Mathematics (Springer, Berlin, 2010); 4th ed. 2010. Corr. 3rd printing 2012 edition (October 31, 2010)
4. F. Harary, *Graph Theory* (1975)
5. D. West, *Introduction to Graph Theory Paperback - 2000, PHI Learning*, 2nd edn. (Prentice Hall, Englewood Cliffs, 2000)

Parallel and Sparse Matrix Computations

5

Abstract

Parallel processing is performed by distributing tasks and/or data of computation to a number of processors. We first briefly review parallel processing concepts in this chapter. Two ways of achieving this goal are shared memory and distributed memory approaches as we describe. We then look at ways of parallelizing matrix multiplications using Python, which is a building block of many matrix and algebraic graph algorithms. The last part of the chapter deals with sparse matrices that have zeros as majority of its elements. We look at ways of representing them in memory and discuss basic operations such as multiplication that make use of sparse matrix property.

5.1 Parallel Processing

A basic computer architecture consists of a processor, a memory system and an input/output system. There is a single flow of instruction in such a system with the processor fetching instructions from memory, decoding and then executing them. A parallel processing system employs a number of such computing systems to solve large and typically data-intensive problems.

5.1.1 Architectures

We can basically have shared memory and distributed memory parallel processing architectures. A shared memory system for parallel processing is characterized by a global memory accessed by all processors mainly for interprocessor communications and synchronization. This memory is in close vicinity to all processors, thus providing fast access but it has to be protected against concurrent accesses. Also, shared memory is a bottleneck when the number of processors is large.

© Springer Nature Switzerland AG 2021

K. Erciyes, *Algebraic Graph Algorithms*, Undergraduate Topics in Computer Science,
https://doi.org/10.1007/978-3-030-87886-3_5

On the other hand, each processor has its local memory in a distributed memory system and main method of communication is via *message-passing*. This architecture can be built by connecting off-the-shelf computers over a communication network.

5.1.2 Basic Communications

A very basic mode of communication between the processes of a parallel computing system is point-to-point transfer of data using the primitives *send* and *receive*. The *send* routine may be implemented as *blocking* meaning the invoker of the procedure is blocked until a response from the receiver is obtained, or *non-blocking* which means the invoker continues execution after calling the procedure. The blocking *receive* routine blocks the caller until a message is received and a non-blocking one simply checks whether a message is received and continues irrespective of the existence of a message. A non-blocking *send* and a blocking *receive* is commonly used in a distributed memory system since guaranteed delivery of a message is usually provided by the network protocols and the next sequence of events by the receiver often depends on the context of the received message.

Many parallel program implementations require participation of a group of processes necessitating higher level methods of communication as follows. Commonly, these operations make use of the underlying parallel computing architecture. We will assume there are p processors (or processes) in the system.

- *One-to-All Broadcast and All-to-One Reduction*: A process having data m sends it to all other processes in the broadcast operation and each process has a local copy of m in the end. In all-to-one reduction, data from each process is collected in a single process possibly using some operation such as addition or finding the maximum value of data.
- *All-to-All Broadcast and All-to-All Reduction*: Each process sends its data to all other processes in all-to-all broadcast. If a process P_i has data m_i to send, each process has $m_0, .., m_{p-1}$ at the end of this operation. All-to-all reduction works similar to all-to-all broadcast but in reverse direction.
- *Scatter and Gather*: In the *scatter* operation, a process sends a personalized message to all other processes and in the *gather* operation, a single process gathers a unique message from all other processes.
- *All-to-All Personalized Communication*: This form of communication results in each process sending a unique message to all other processes. Thus, it is a generalized form of scatter.

5.1.3 Performance

The speedup of a parallel algorithm provides comparison of a parallel algorithm running time T_p to that of the best sequential algorithm T_s as below.

$$S = \frac{T_s}{T_p}$$

Speedup is not equal to the number of processors used simply because interprocess communication time is included in parallel processing time. Thus, T_p increases with the number of processors used. The efficiency E of a parallel algorithm needs to consider the number of processors p used as follows:

$$E = \frac{S}{p}$$

A *scalable* parallel algorithm has almost a constant efficiency when the number of processors and the size of the problem is increased.

5.1.4   Parallel Matrix Computations

Matrix-vector multiplication and matrix-matrix multiplication are two of the most commonly used basic building block algorithms of higher level tasks. There are various algorithms to implement these operations in parallel and we will consider basic approaches that distribute data in a distributed memory computer system.

5.1.4.1   Matrix-Vector Multiplication
Sequential matrix-vector multiplication of a matrix $n \times n$ matrix A with a vector x of size n requires n^2 multiplications and additions. Let us consider this multiplication in a parallel processing system with p processors. We can distribute n/p rows of matrix A to each processor P_i and broadcast vector x to all processors. Each processor then forms the product of its associated rows in parallel as shown in Fig. 5.1a in this *rowwise 1D partitioning*.

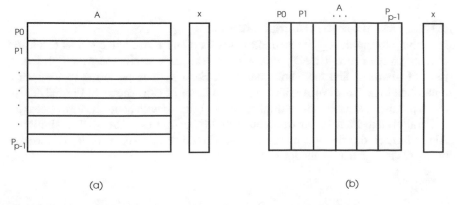

Fig. 5.1 Rowwise and columnwise partitioning for matrix-vector multiplication

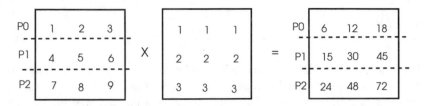

Fig. 5.2 Rowwise partitioning for matrix-matrix multiplication

This parallel operation to form $Ax = B$ may be realized by a master process that distributes rows of A using scatter operation, broadcasting vector x to all and then gathering the results as shown in Algorithm 5.1.

Algorithm 5.1 *Matrix-Vector Multiplication*

1: **Input** : matrix $A_{m \times n}$, vector x_n
2: **Output** : product $c \in \mathbb{R}$
3: **if** *me == master* **then**
4:      **scatter** rows size n/p of A to all processes
5:      **broadcast** vector x to all processes
6:      **gather** rows of c from all processes
7: **else**
8:      **receive** my_A_rows from master
9:      **compute** my_c_rows
10:     **send** my_c_rows to master
11: **end if**

Columnwise 1D partitioning of matrix A involves sending n/p columns of matrix A to each processor p_i and broadcast vector x as in rowwise 1D partitioning which is shown in Fig. 5.1b.

5.1.4.2 Matrix-Matrix Multiplication

In forming the product C of two $n \times n$ matrices A and B, we will consider extension of matrix vector rowwise 1D partitioning. In this case, each processor P_i has n/p rows of matrix A and the whole matrix B. Each P_i then performs vector-matrix multiplication of row i of A with B to obtain row i of C as shown in Fig. 5.2. There exists parallel matrix-matrix multiplication algorithms with better performances [5].

Parallel algorithm for matrix-matrix multiplication $C = A \times B$ is similar to Algorithm 5.1, this time the master process should broadcast matrix B to all worker processes and gather results for matrix C.

5.2   Parallel Computations with Python

There exists various parallel processing methods which can be used with Python. Three libraries for shared memory parallel processing are *thread*, *threading* and *multiprocessing*. Although thread-based parallel processing is faster than process-based parallel processing, these libraries are commonly preferred for input/output intensive applications. Thus, we will review *multiprocessing* library which is process-based and is convenient for data intensive applications.

5.2.1   Process-Based Model with *Multiprocessing*

The library associated for parallel process-based models in Python is called *multiprocessing*. The *Pool* object in this module provides data parallelism by distributing the input data to a number of specified processes. The following example shows first to output the number of cores in the running environment using the method *cpu_count* and how to use *Pool* object to spawn 5 processes each of which prints the square of the input and returns it [3]. Note that the *map* method on *Pool* object *p* distributes integers 1,...,5 to 5 processes to perform function *sq* for square calculation.

```
1   from multiprocessing import Pool
2   import multiprocessing
3
4   def sq(x):
5       return x*x
6
7   if __name__ == '__main__':
8       print(multiprocessing.cpu_count())
9       p = Pool(5)
10      print(p.map(sq, [1, 2, 3, 4, 5]))
11  >>>
12  2
13  [1, 4, 9, 16, 25]
```

Here, we see that the system has a dual-core processor. An alternative way of achieving parallel processing using the *multiprocessing* module is by using the *Process* object. A process is spawned by calling *Process* object of multiprocessing library with the name of the process and possible arguments. The following example shows how to create 5 worker processes by a master process. The identifier *p* of a process returned after spawning is stored in the list *procs* as we will use this parameter to start a process by the *start* method and wait for the process by the *join* method. Note that we declare a process as an ordinary Python function as in *Pool* object implementation.

```
1    import multiprocessing as mp
2
3    def worker(me):
4        print ("Worker", me)
5        return
6
7    if __name__ == '__main__':
8        procs = []
9        print("master")
10       for i in range(5):
11           p = mp.Process(target=worker,args=(i,))
12           procs.append(p)
13           p.start()
14       for p in procs:
15           p.join()
16   >>>
17   master
18   ('Worker', 0)
19   ('Worker', 3)
20   ('Worker', 2)
21   ('Worker', 4)
22   ('Worker', 1)
```

5.2.1.1   Parallel Matrix-Matrix Multiplication

We will now implement parallel matrix multiplication by row-partitioning using the *Process* object of *multiprocessing* module. The input matrices A and B are small 3×3 matrices and each of 3 processes receives its row and matrix B from the master process, multiplies its row with matrix B to form the row of product matrix C. A worker then sends its associated row of matrix C to the master. We need to consider the following for this algorithm: when a process is spawned using the *Process* module, it has all of the data of the parent process but the modifications it makes are not reflected in the parent process. In order to have the product matrix C shared between the processes, we need to use the *Manager* object of the multiprocessing module and define C as an array to be shared from this object. Each worker process performs multiplication and stores the result in its row entry of matrix C.

```
1    import multiprocessing as mp
2    import numpy as np
3
4    A = np.array([[1,2,3],[4,5,6],[7,8,9]])
5    B = np.array([[1,2,3],[1,2,3],[1,2,3]])
6
7    def worker(me,C,l):
8        l.acquire()
9        Y = np.dot(A[me,:],B)
10       print("me and my product",me, Y)
11       C.append(Y)
12       l.release()
13       return Y
14
```

```
15  if __name__ == '__main__':
16      mgr = mp.Manager()
17      C = mgr.list()
18      lock = mp.Lock()
19      procs = []
20      for i in range(3):
21          p = mp.Process(target=worker,args=(i,C,lock))
22          procs.append(p)
23          p.start()
24      for proc in procs:
25          proc.join()
26      print(np.array(C))
27  >>>
28  ('me and my product', 0, array([ 6, 12, 18]))
29  ('me and my product', 2, array([24, 48, 72]))
30  ('me and my product', 1, array([15, 30, 45]))
31  [[ 6 12 18]
32   [24 48 72]
33   [15 30 45]]
```

5.2.1.2 Parallel Calculation of π

Another way of sharing data between processes of the multiprocessing module is using the *Queue* class. This class provides *put* method for storing data and *get* method for retrieving data between processes. Typically, the main process that spawns all other processes creates a *Queue* object and passes it as a parameter to the process that the result is expected. The spawned process puts the data to be returned in the queue which is taken by the *get* method.

Let us consider the example of calculating the number π using integration to show the *Queue* class for sharing data. The number π can be approximated by the area under the curve between 0 and 1 values of variable x as follows,

$$\int_0^1 \frac{4}{1+x^2} \approx \pi$$

Let us first check whether this approximation holds and draw this function using Python library *scipy* class *integrate* to compute this integral and *matplotlib* library to plot the function.

```
1   import matplotlib.pyplot as plt
2   import numpy as np
3   import scipy.integrate as integrate
4
5   result = integrate.quad(lambda x: 4/(1+x**2), 0, 1)
6   print (result[0])
7
8   x = np.linspace(0,1,10)
9   plt.plot(x, 4/(1+x**2))
10  plt.show()
11  >>>
12  3.141592653589793
```

Fig. 5.3 The π curve

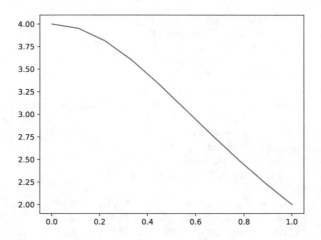

The proposed evaluation of π provides a very good approximation. The plot of this function against x values between 0 and 1 is depicted in Fig. 5.3.

We will now calculate π by finding the area under this curve which is equal to the integration. Each process spawned has a segment of the curve to work and divides this segment into a number of thin slices to calculate. The main process sets the total number of slices *nslices* to one million, the number of processes *nprocs* to 4 and sends these arguments to each process it spawns together with the queue object *que* to retrieve data. Each worker process *calc_pi* then calculates its own area by dividing its portion of the curve into thinner slices using *nsteps*, adds the areas under these slices and puts its resulting total area to *que*. The main process gathers all these results, sums them and prints the output which is quite close to actual *pi* value as displayed.

```
1   mport multiprocessing as mp
2
3   def calc_pi(rank, nslices, nprocs,que):
4       partial_pi = 0.0
5       dx = 1.0 / nslices
6       n = nsteps / nprocs
7       for i in range(1,n):
8           x = (1.0/nprocs * rank) + (dx * i)
9           y = 4.0 / (1.0 + (x * x))
10          area = dx * y
11          partial_pi += area
12      que.put(partial_pi)
13
14  if __name__ == '__main__':
15      nslices = 1000000
16      nprocs = 4
17      que = mp.Queue()
18      procs = []
19      for i in range(nprocs):
20          p = mp.Process(target=calc_pi, args=(i,nslices,
```

```
21          nprocs,que,))
22          procs.append(p)
23
24      for p in procs:
25          p.start()
26
27      pi_total = 0.0
28      for p in procs:
29          p.join()
30          pi_total = pi_total + que.get()
31      print(pi_total)
32  >>>
33  3.14158012888
```

5.2.2   Message Passing Interface *mpi4py*

The message passing interface (MPI) is a standard for message passing for parallel programs that run on distributed memory computers [6]. It also has shared memory or hybrid implementations. MPI provides routines for sending and receiving point-to-point, multicast and broadcast messages between distributed programs. The message passing interface (MPI) library provides a rich set of routines for parallel processing over a network. Each program running MPI belongs to a *communicator* and each program inside a communicator has a unique identifier called its *rank*. An MPI program at a node of a computer network starts by initializing the environment,

The MPI implementation for Python is called $mpi4py$ which contains most of the routines in the MPI standard. We import MPI module from $mpi4py$ library as the first line in the following Python code. The running program then builds a *communicator* which is basically the label of a data structure identifying the running processes. It then finds its *rank* which is its identifier and the *size* of the communicator which is the number of processes in the communicator using the related system calls from the MPI module.

```
1   # ranks.py
2   from mpi4py import MPI
3
4   comm = MPI.COMM_WORLD
5   rank = comm.Get_rank() # find rank
6   size = comm.Get_size() # find size
7   print(rank, size)
8   >>>
9   (2, 4)
10  (3, 4)
11  (0, 4)
12  (1, 4)
```

The following command entered from console will cause the displayed output. Note that the number of processes is passed as input to the program, to be stored as size of the communicator which is 4 in this case.

```
mpirun -n 4 python ranks.py
```

5.2.2.1 Point-to-Point Communication

Two routines for point to point communication in *mpi4py* are *send* and *receive* routines as below. These are blocking operations, that is, the sender gets blocked until the receiver receives the message and the receiver gets blocked until a message arrives.

- *comm.send(data, destination, tag)*: Data is the data to be sent, destination is the rank of the receiver and tag is the type of message. Having tag field provides sending of different types of messages over the network.
- *comm.recv(source, tag, status)*: Source is the rank of the sender or it can be MPI.ANY_SOURCE to receive any message sent by any node. The tag field is the type of message as in send and the status field is used to obtain in information about the received message.

The following Python code demonstrates sending a message from rank 0 process to rank 1 process which displays it. Note that we can omit some of the parameters in *send* and *receive* routines.

```
1   # data.py
2   from mpi4py import MPI
3
4   comm = MPI.COMM_WORLD
5   rank = comm.Get_rank()
6   size = comm.Get_size()
7
8   # master process
9   if rank == 0:
10      data = [[1,2,3],[4,5,6],[7,8,9]]
11      for i in range(1, size):
12          comm.send(data[i-1], dest=i, tag=i)
13          print('Process {} sent data:'.format(rank), data[i-1])
14
15  # worker processes
16  else:
17      data = comm.recv(source=0, tag=rank)
18      print('Process {} received data:'.format(rank),data)
19  >>>
20  ('Process 0 sent data:', [1, 2, 3])
21  ('Process 0 sent data:', [4, 5, 6])
22  ('Process 0 sent data:', [7, 8, 9])
23  ('Process 2 received data:', [4, 5, 6])
```

```
24  ('Process 1 received data:', [1, 2, 3])
25  ('Process 3 received data:', [7, 8, 9])
```

More advanced message sending and receiving operations are specified as follows.

- *comm.bcast(data, root=0)*: Data in message is broadcast to all members of the communicator. Each process calls the same procedure.
- *comm.scatter(data, root=0)*: Data consists of partitions and each partition is sent to a different process.
- *comm.gather(data, root=0)*: This routine functions as the reverse operation of the scatter in which each process receives its related partition of data. Broadcast operation is shown in the Python code below.

```
1   # bcast.py
2   from mpi4py import MPI
3   comm = MPI.COMM_WORLD
4   rank = comm.Get_rank()
5
6   if rank == 0:
7       data = [1,2,3,4,5]
8   else:
9       data = None
10  data = comm.bcast(data, root=0)
11  print(rank,data)
12  >>>
13  (0, [1, 2, 3, 4, 5])
14  (1, [1, 2, 3, 4, 5])
15  (3, [1, 2, 3, 4, 5])
16  (2, [1, 2, 3, 4, 5])
```

5.2.2.2 Parallel Matrix Multiplication

We will first implement parallel matrix multiplication $C = A \times B$ by row partitioning matrix A and sending each process its partition and the whole of matrix B. Each process performs its own multiplication and sends the partial product to the master process which collects all results and then prints the product matrix C, we have 4 processes including the master process. The output matrix C displayed is the product of matrices A and B.

```
1   # mult.py
2   from mpi4py import MPI
3   import numpy as np
4
5   comm = MPI.COMM_WORLD
6   rank = comm.Get_rank()
7   size = comm.Get_size()
8
9   # master process
```

```
10   if rank == 0:
11       A = np.array([[1,2,3],[4,5,6],[7,8,9]])
12       B = np.array([[1,2,3],[1,2,3],[1,2,3]])
13       C = np.zeros([3,3])
14
15       for i in range(1, size):
16           comm.send(A[i-1], dest=i, tag=i)
17           comm.send(B, dest=i, tag=i)
18       for i in range(1, size):
19           C[i-1,:] = comm.recv(source=i, tag=rank)
20       print(C)
21
22   # worker processes
23   else:
24       A = comm.recv(source=0, tag=rank)
25       B = comm.recv(source=0, tag=rank)
26       M = np.dot(A,B)
27       comm.send(M, dest=0, tag=0)
28   >>>
29   [[ 6. 12. 18.]
30    [15. 30. 45.]
31    [24. 48. 72.]]
```

We can improve this implementation by using higher level system calls of MPI which are *broadcast*, *scatter* and *gather* operations. In this implementation, we have also the master process involved in computation as in the code below.

```
1    //mult.py
2    from mpi4py import MPI import numpy as np
3
4    comm = MPI.COMM_WORLD rank = comm.Get_rank() n = comm.Get_size()
5
6    if rank == 0:
7        A = [[1,2,3],[4,5,6],[7,8,9]]
8        B = [[1,2,3],[1,2,3],[1,2,3]]
9    else:
10       A, B  = None, None
11
12   my_data = [[-1]*n] B = comm.bcast(B, root=0) # root broadcasts B
13   my_data = comm.scatter(A, root=0)  # root scatters A C =
14   np.dot(my_data,B)  # each process finds local product D =
15   comm.gather(C,root=0) # root gathers products if rank == 0:
16   # root prints result
17       print(np.asarray(D))
18   >>>
19   [[ 6 12 18]
20    [15 30 45]
21    [24 48 72]]
```

5.2.2.3   Parallel Calculation of π

We will now implement the same algorithm to calculate *pi* using mpi4py similar to [2]. The worker process finds the area under the function $\int_0^1 \frac{4}{1+x^2}$ for its associated part as before. The number of processes is broadcast to all workers which can find their portions and then do the calculations using the function *calc_pi*. The root process then collects all results from the workers by adding them using the *reduce* operation and prints the result. Note that we can specify what to do with the collected data in *reduce* operation.

```python
# pi.py
from mpi4py import MPI
import math

def calc_pi(nslices, rank, nprocs):
    k = nslices/nprocs
    dx = 1.0 / nslices
    f = 0.0
    for i in range(0, k, 1):
        x =  dx * (i + 0.5 + rank * k) # find x value
        f += 4.0 / (1.0 + x**2)        # add y values
    return f * dx                      # calculate area

comm = MPI.COMM_WORLD
nprocs = comm.Get_size()
rank = comm.Get_rank()
if rank == 0:
    nslices = 1000                     # total number of slices
else:
    nslices = None

nslices = comm.bcast(nslices, root=0)  # broadcast to workers

pi_worker = calc_pi(nslices, rank, nprocs) # calculate local
pi = comm.reduce(pi_worker, op=MPI.SUM, root=0) # collect

if rank == 0:
    error = abs(pi - math.pi)
    print ("pi: %.16f, ""error: %.16f" % (pi, error))
>>>
pi is approximately 3.1415927369231262,
error is 0.0000000833333331
```

5.3   Sparse Matrix Algorithms

A sparse matrix consists of mostly zero elements. Formally, the number of nonzero elements (NNZ) of an $n \times n$ spares matrix is $O(n)$ and the *sparsity* of a matrix is the ratio of the number of its zero elements to the total number of its elements. As an example, a 4×6 sparse matrix with 24 elements is given below. Since the number of zeros is 13, sparsity of this matrix is $17/24 \approx 0.708$.

$$\begin{bmatrix} 1 & 0 & 0 & 1 & 0 & 0 \\ 0 & 2 & 0 & 0 & 1 & 0 \\ 1 & 0 & 0 & 0 & 0 & 0 \\ 0 & 0 & 1 & 0 & 2 & 0 \end{bmatrix}$$

Sparse matrices are commonly used in machine learning, natural language processing and various scientific problems. Most of the large matrices are sparse, thus, efficient methods for storage and computation of these matrices are needed.

5.3.1 Sparsity Calculation

We can calculate sparsity of a sparse matrix as in the below Python code by finding the percentage of the number of non-zero elements in the matrix and subtracting this value from 1 as shown in the below Python code for the above example which finds the manually calculated sparsity.

```
import numpy as np

# create dense matrix
A = np.array([[1,0,0,1,0,0], [0,2,0,0,1,0], [1,0,0,0,0,0],
               [0,0,1,0,2,0]])
print(A)

# calculate sparsity
sparsity = 1.0 - np.count_nonzero(A) / A.size
print(sparsity)
>>>
[[1 0 0 1 0 0]
 [0 2 0 0 1 0]
 [1 0 0 0 0 0]
 [0 0 1 0 2 0]]
0.7083333333333333
```

5.3.2 Sparse Matrix Representations

Since the number of non-zero elements of a sparse matrix is small compared to its total number of elements, storing a sparse matrix as a dense matrix is a waste of memory. Thus, various methods to store sparse matrices have been devised. Computations using sparse matrices as dense matrices also means wasting computational power. Two basic methods for sparse matrix representations are array representation and linked list representation.

The array representation makes use of a triplet as (row, column, value) of non-zero elements of the matrix. The compressed sparse row (CSR) and compressed sparse column (CSC) are two basic formats commonly used for the storage of sparse matrices using arrays. The CSR format stores the triplets rowwise and the CSC format as columnwise.

A sparse matrix can be created by the method *csr_matrix* from the Python library *scipy*. The following code segment demonstrates specifying non-zero elements of a sparse matrix as arrays of rows, columns and data. For example, row[0, col[0] and data[0] corresponds to value data[0] at [0,0] element of the matrix. These arrays together with the dimension of the matrix are given as arguments to the method *csr_matrix* which forms the sparse matrix and converts it to array *S* for display.

```
import numpy as np
from scipy.sparse import csr_matrix

row  = np.array([0, 0, 1, 1, 2, 3])
col  = np.array([0, 3, 2, 5, 3, 5])
data = np.array([1, 4, 2, 1, 3, 5])

S = csr_matrix((data, (row, col)),
                        shape = (4, 6)).toarray()
print(S)
>>>
[[1 0 0 4 0 0]
 [0 0 2 0 0 1]
 [0 0 0 3 0 0]
 [0 0 0 0 0 5]]
```

Python also provides methods to convert a given dense matrix to a sparse matrix in the required format and convert the sparse matrix to a dense matrix as shown in the below code. The dense matrix *A* is converted to CSR format by the *csr_matrix* method and it is converted back to dense form by the *todense* method.

```
from numpy import array
from scipy.sparse import csr_matrix

# create dense matrix
A = array([[1,0,0,0,4,0], [0,0,2,0,0,1], [0,0,0,3,0,0],
           [0,0,0,0,0,5]])
print("A: \n",A)
# convert to sparse matrix (CSR method)
S = csr_matrix(A)
print("S: \n",S)
# reconstruct dense matrix
B = S.todense()
print("B: \n",B)
>>>
A:
 [[1 0 0 0 4 0]
```

```
 [0 0 2 0 0 1]
 [0 0 0 3 0 0]
 [0 0 0 0 0 5]]
S:
   (0, 0)            1
   (0, 4)            4
   (1, 2)            2
   (1, 5)            1
   (2, 3)            3
   (3, 5)            5
B:
 [[1 0 0 0 4 0]
 [0 0 2 0 0 1]
 [0 0 0 3 0 0]
 [0 0 0 0 0 5]]
```

Other types of sparse matrix storage supported in Python *scipy* class are as follows.

- *bsr_matrix*: Block Sparse Row (BSR) format is used when sparse matrix has some dense submatrices.
- *lil_matrix*: List of Lists (LIL) format is basically a linked list of nonzero elements of the matrix where each such element points to the next rowwise nonzero element.
- *dok_matrix*: Dictionary of Keys (DOK) format allows storing nonzero elements as *{row,column}*:value (*key-value*) pairs of a dictionary.
- *coo_matrix*: COOrdinate format is used for fast sparse matrix creation. The formed matrix can then be converted to CSR or CSC formats for fast arithmetic and matrix-vector operations.
- *dia_matrix*: DIAgonal format enables storing the nonzero elements along the diagonal of a matrix.

5.3.3 Sparse Matrix Multiplication

Sparse matrix-vector multiplication in Python is similar to *numpy.dot* operation, however the provided @ operator is used. Let us consider multiplying a 5 by 5 sparse matrix with a 5 by 3 sparse matrix shown below.

$$
\begin{bmatrix} 0 & 1 & 0 & 0 & 0 \\ 0 & 0 & 0 & 1 & 0 \\ 1 & 0 & 0 & 0 & 1 \\ 1 & 0 & 0 & 0 & 0 \\ 0 & 1 & 0 & 0 & 0 \end{bmatrix} \times \begin{bmatrix} 1 & 0 & 0 \\ 0 & 0 & 1 \\ 0 & 0 & 0 \\ 0 & 1 & 0 \\ 1 & 0 & 0 \end{bmatrix}
$$

The Python code to implement multiplication of these two sparse matrices is given below. We first specify the matrices $S1$ and $S2$ in CSR format and then multiply them and print the product in both CSR format and as dense matrix.

```
import numpy as np
from scipy.sparse import csr_matrix

row1  = np.array([0, 1, 2, 2, 3, 4])
col1  = np.array([1, 3, 0, 4, 0, 1])
data1 = np.array([1, 1, 1, 1, 1, 1])
row2 = np.array([0, 1, 3, 4])
col2  = np.array([0, 2, 1, 0])
data2 = np.array([1, 1, 1, 1])

S1 = csr_matrix((data1, (row1, col1)), shape = (5, 5))
S2 = csr_matrix((data2, (row2, col2)), shape = (5, 3))
S = S1 @ S2
print("S in CSR format \n", S, "\n S as dense \n ", S.toarray())
>>>
S in CSR format
    (0, 2)          1
  (1, 1)          1
  (2, 0)          2
  (3, 0)          1
  (4, 2)          1
 S as dense
  [[0 0 1]
 [0 1 0]
 [2 0 0]
 [1 0 0]
 [0 0 1]]
```

5.4   Chapter Notes

Parallel computations may be carried out in shared memory or distributed memory models. Processes communicate using shared memory in the former which requires protection of memory and synchronization. The main method of communication in distributed memory model is by message passing.

Matrix multiplication is a fundamental building block operation involved in various matrix computations used in algebraic graph algorithms. Matrix-matrix multiplication can be performed by rowwise or columnwise partitioning, block partitioning and cyclic partitioning. We reviewed basic rowwise and columnwise partitioning.

Various modules for Python provide parallel computation. Out of these modules, we described the shared memory module *multiprocessing* which yields routines for process spawning and synchronization. The distributed memory parallel processing standard MPI is widely used in practice and Python module *mpi4py* provides most of the system calls specified in this module. We showed various parallel matrix computations using this module.

We have barely touched the surface of parallel processing in this chapter. However, parallel matrix computations with Python with focus on matrix multiplications was reviewed in detail since we will be using matrix computations frequently to solve many algebraic graph algorithms. A thorough review of parallel processing in general is covered in [5].

Programming Projects

1. Write a Python program using the *multiprocessing* method and the *Pool* object that spawns 10 processes with inputs 1,...,10 and each process calculates the sum of integers up to and including its input and returns this sum.
2. Provide a Python program that multiplies to $n \times n$ matrices using columnwise 1D partitioning.
3. Find the number π by finding the area under the curve $\int_0^1 \frac{4}{1+x^2}$ using *Pool* class of *multiprocessing* module.
4. Write a Python program that converts a sparse matrix and a vector in dense form to CSC format, multiplies them and prints the product.

References

1. L. Dalcin, R. Paz, M. Storti, MPI for Python. J. Parallel Distrib. Comput. **65**(9), 1108–1115 (2005)
2. L. Dalcin, R. Paz, M. Storti, J.D. Elia, MPI for Python: performance improvements and MPI-2 extensions. J. Parallel Distrib. Comput. **68**, 655–662 (2008)
3. https://docs.python.org/3/library/multiprocessing.html
4. M.R. Garey, D.S. Johnson, *Computers and Intractability* (W.H. Freeman, New York, 1979)
5. A. Gupta, G. Karypis, V. Kumar, A. Grama, *Introduction to Parallel Computing, Design and Analysis of Algorithms*. Pearson College Div; Subsequent edition (January 1, 2003) (2003)
6. http://www.mcs.anl.gov/research/projects/mpi

Part II
Graph Algorithms

Trees

6

Abstract

A tree is graph that does not contain any cycles. A tree may be used to model many real network structures such as the administration of an organization. We first provide a method to construct a spanning tree in this chapter. We then review few algorithms to construct a minimum spanning tree of a weighted graph.

6.1 Introduction

A *tree* is a connected acyclic graph. Trees have numerous applications; for example, the structure of an organization may be shown by a tree with the manager at the top of the tree and in biology, phylogenetic trees depict the relationship between species. A tree $T = (V', E')$ of a graph $G = (V, E)$ is an acyclic subgraph of G. A *spanning tree* of a graph G is a tree of G that contains all vertices of G. In a weighted graph G where edges have weights associated with them, a *minimum spanning tree* is a spanning tree of G that has the total minimum weight of edges among all spanning trees of G. The following statements equally define a tree:

- A tree with n nodes has exactly $n - 1$ edges.
- A connected undirected graph with n nodes is a tree if it has $n - 1$ edges.
- An undirected graph is a tree if and only if there is a unique path between any pair of its nodes.

We start with simple spanning tree implementations in this chapter and continue with algebraic algorithms to form a minimum spanning tree. We will also sketch parallel formation of algorithms for exemplary cases.

© Springer Nature Switzerland AG 2021   89
K. Erciyes, *Algebraic Graph Algorithms*, Undergraduate Topics in Computer Science,
https://doi.org/10.1007/978-3-030-87886-3_6

6.2 Spanning Tree Construction

A spanning tree of a graph is a tree that contains all of the vertices of the graph as
stated. Let us first define this structure formally.

Definition 6.1 (*spanning tree*) Given a graph $G = (V, E)$, a subgraph $G' = (V, E')$
of G where $E' \subseteq E$ is called an *induced subgraph* of G. A spanning tree T of a
graph G is an induced subgraph of G which is a tree containing all vertices of G.
Alternatively, a spanning tree $T = (V, E')$ of a graph $G = (V, E)$ is a tree and
subgraph of G that covers all of the vertices of G.

Building a spanning tree out of a graph has many applications; we will first look
at building an arbitrary spanning tree of a graph in this part. A simple way to build a
spanning tree T is to start with an arbitrary vertex $u \in G$ and pick an outgoing edge
(u, v) from u, include (u, v) in the tree. The next iteration will select any outgoing
edge (u, w) or (v, z) from u or v in the set $T' = \{u, v\}$, and include the edge (u, w)
or (v, z) in T' and so on until all vertices are processed. Algorithm 6.1 depicts the
operation of this algorithm.

The simple graph of Fig. 6.1 has 4 nodes and implementing this algorithm results
in the spanning tree of (d) in three iterations.

Fig. 6.1 Construction of a
spanning tree of a simple
graph. The starting node is 1
and each outgoing edge
included in the tree is shown
in dashes

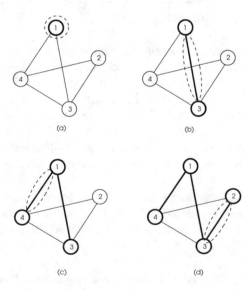

Algorithm 6.1 *Spanning Tree Algorithm*

1: **Input** : $G = (V, E)$ ▷ unweighted connected graph G
2: **Output** : $T = (V, E_T)$ ▷ spanning tree of G
3: $V' \leftarrow \{s\}$
4: $T \leftarrow \varnothing$
5: **while** $V' \neq V$ **do**   ▷ continue until all vertices are visited
6:     select an arbitrary edge (u, v) such that $u \in T$ and $v \in G \setminus T$
7:     $V' \leftarrow V' \cup \{v\}$
8:     $E_T \leftarrow E_T \cup \{(u, v)\}$
9: **end while**

Python Implementation

Let us consider an algebraic version of this algorithm with the graph G represented by an adjacency matrix A. Let the vertices of the graph be numbered from 0 to $n - 1$ and $A[i, j] = 1$ if there is an edge between vertices i and j as in the usual notation. We will implement Algorithm 6.1 in Python with the following data structures:

- *edges*: A dictionary which has all edges incident to a vertex at each entry.
- *P*: A dictionary which shows parent of a vertex in the final spanning tree.
- *V*: A vector that shows whether a vertex is in the spanning tree or not.
- *vis*: A list that contains visited vertices.

The algorithm starts by filling all the entries in *edges* in lines and initializing dictionary *P*, vector *V* and list *vis*. The first vertex that is visited is vertex 0 which is appended to *vis* and its parent is set to itself. The main body of the algorithm searches for an outgoing edge from the set of current tree vertices by testing incident edges of vertices in *vis*. We need to check whether any outgoing edge makes a cycle with existing edges in the spanning tree. For this reason, the other end of an edge from a vertex in tree is tested, if it is already in tree with V having a value 1 for that vertex, other vertices are checked until an outgoing edge is found. The newly found vertex is then appended to vis, the edge that is incident to it is removed from edges to discard it from further search and the inner loops are abandoned. This process continues until $n - 1$ edges are contained in the tree. The returned output from this function is the P dictionary which holds parents for every vertex in the graph.

```
1   ####################################################################
2   #                  Spanning Tree Algorithm                         #
3   ####################################################################
4
5   import numpy as np
6
7   def Spanning_Tree(A):
8
9     n=len(A)
```

```
10    edges = {edges: [] for edges in range(n)} # edge list
11    P = {P: [] for P in range(n)} # parent list
12    V = np.zeros(n)              # vertices in ST
13    vis = []                     # visited
14
15    for i in range(0,n):    # initialize edge list
16       for j in range (i+1,n):
17             if A[i,j] != 0:
18                 edges[i].append((i,j))
19    V[0] = 1
20    P[0] = 0
21    vis.append(0)
22
23    count = 0
24    while count < n-1:    # iterate until list is empty
25       for i in range(len(vis)):
26             for j in range(len(edges[i])):
27                 x = edges[i][j][1]
28                 if V[x] == 0:
29                     V[x] = 1
30                     P[x] = i
31                     vis.append(x)
32                     del edges[i][j]
33                     count = count +1
34                     break
35             if V[x] == 0:
36                     break
37    return P
38
39 if __name__ == '__main__':
40    B = np.array([[0,1,0,0,0,0,1],
41                  [1,0,1,0,0,1,1],
42                  [0,1,0,1,0,1,0],
43                  [0,0,1,0,1,1,0],
44                  [0,0,0,1,0,1,0],
45                  [0,1,1,1,1,0,1],
46                  [1,1,0,0,0,1,0]])
47
48    T = Spanning_Tree(B)
49    print ("Spanning Tree", T)
50 >>>
51 Spanning Tree: {0: 0, 1: 0, 2: 1, 3: 2, 4: 3, 5: 1, 6: 0}
```

Running of this algorithm in the sample graph of Fig. 6.2 resulted in the spanning tree displayed. The inner most loop of the algorithm runs $\Delta(G)$ times and the outer loops each take $O(n)$ time resulting in $O(\Delta n^2)$ time for this algorithm.

Fig. 6.2 A sample graph to test Python spanning tree algorithm. Tree edges are shown in bold with arrows pointing to the parent

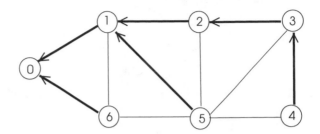

6.3 Minimum Spanning Trees

A minimum spanning tree (MST) of a weighted graph $G = (V, E, w)$ is a spanning tree of G with a total minimum weight of edges among all spanning trees of G.

Definition 6.2 (*minimum spanning tree*) Let $w(e)$ denote the weight of an edge and $w(T)$ denote the total weight of edges in any spanning tree T of a weighted graph $G = (V, E, w)$ as below,

$$w(T) = \sum_{(e) \in T} w(e)$$

The MST of G is the tree T with minimum $w(T)$ value among all spanning trees of G.

It can be shown that a graph with distinct weight values of edges has a unique MST [1]. MSTs have numerous applications, for example, building a communication network using a minimum length of cables requires building an MST. The classical algorithms to build an MST are due to Jarnik-Prim, Kruskal and Boruvka as described in the following sections.

6.3.1 Jarnik-Prim Algorithm

The Jarnik-Prim (JP) algorithm is similar to the spanning tree algorithm with one major difference, we have a weighted graph and search for the minimum weight outgoing edge (MWOE) at each iteration while enlarging the current MST fragment. The classical algorithm is depicted in Algorithm 6.2. It can be shown that this algorithm is correct and runs in $O(m \log n)$ time [1].

Running of this algorithm in a small graph with 6 nodes is shown in Fig. 6.3.

6.3.1.1 The Algebraic Algorithm

We will describe the implementation of JP algorithm in an array based language like MATLAB. The distance matrix D of the graph is input to the algorithm and the total weight w_t and the parent of a node in the MST is the output of this algorithm. Let S be the current vertices assigned to the MST and s be a binary vector that denotes whether

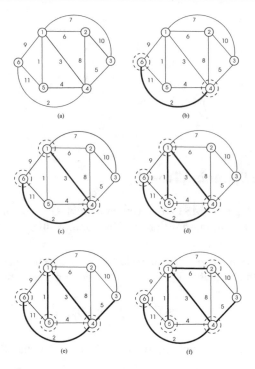

Fig. 6.3 Running of JP Algorithm in a sample graph

Algorithm 6.2 *JP_MST*

1: **Input** : $G = (V, E, w)$
2: **Output** : MST $T(V_T, E_T)$ of G
3: $V_T \leftarrow \{s\}$
4: $T \leftarrow \emptyset$
5: **while** $V_T \neq V$ **do**   ▷ continue until all vertices are visited
6:    select the edge (u, v) with minimal weight such that $u \in T$ and $v \in G \setminus T$
7:    $V_T \leftarrow V_T \cup \{v\}$
8:    $E_T \leftarrow E_T \cup \{(u, v)\}$
9: **end while**

---

a vertex in S or not. We also have another vector **d** that shows the best distance of a node to anny node in S, the entry for a node $u \in S$ is 0. At each iteration, we need to find the MWOE (u, v) from S with $u \in S$ and $V \setminus S$ and include v in S as shown in Algorithm 6.3. The vector π holds the parent of an assigned node in the MST.

The vector s is initialized to ∞ and any node that is included in the MST is assigned a value ∞ in this vector to exclude it from further calculations. The *argmin* call at line 9 finds the index of the smallest element of vector obtained by adding **s** to **d**. This way, the vertex with the minimum value connection to any vertex in S is discovered. In the first run, we are discovering the lightest edge that is incident

Algorithm 6.3 *Algebraic JP Algorithm*

1: **Input** : $G(V, E)$, directed or undirected graph
2: **Output** :
3: s=0;
4: $w_t \leftarrow 0$;
5: s(1)=∞;
6: $\pi \leftarrow \emptyset$;
7: $\mathbf{d} = \mathbf{D}(1, :)$;
8: **while** s $\neq \infty$
9:     i=argmin$\{\mathbf{s} + \mathbf{d}\}$;
10:     $\mathbf{s}(i) = \infty$;
11:     $< w, p >= \mathbf{d}(i)$
12:     $w_t = w_t + \mathbf{d}(u)$;
13:     $\pi(i) = p$
14:     $\mathbf{d} = \mathbf{d}.\min \mathbf{A}(i, :)$;
15:

to vertex 1 as **d** is initialized with the fist row of A. The values of vectors **s** and **d** during first three iterations of this algorithm for the graph of Fig. 6.3 are shown below.

$$
\begin{array}{ccc}
1 & 2 & 3
\end{array}
$$
$$\mathbf{s} = \begin{pmatrix}\infty\ 0\ 0\ 0\ 0\ 0\end{pmatrix} \rightarrow \begin{pmatrix}\infty\ 0\ 0\ 0\ \infty\ 0\end{pmatrix} \rightarrow \begin{pmatrix}\infty\ 0\ 0\ \infty\ \infty\ 0\end{pmatrix}$$

$$\pi = \begin{pmatrix}0\ 0\ 0\ 0\ 0\ 0\end{pmatrix} \rightarrow \begin{pmatrix}0\ 0\ 0\ 0\ 1\ 0\end{pmatrix} \rightarrow \begin{pmatrix}0\ 0\ 0\ 1\ 1\ 0\end{pmatrix}$$

$$\mathbf{d} = \begin{pmatrix}0\ 6\ \infty\ 3\ 1\ 9\end{pmatrix} \rightarrow \begin{pmatrix}0\ 6\ \infty\ 3\ 0\ 9\end{pmatrix} \rightarrow \begin{pmatrix}0\ 6\ 5\ 0\ 0\ 2\end{pmatrix}$$

The final three iterations below provides the needed result.

$$
\begin{array}{ccc}
4 & 5 & 6
\end{array}
$$
$$\mathbf{s} = \begin{pmatrix}\infty\ 0\ 0\ \infty\ \infty\ \infty\end{pmatrix} \rightarrow \begin{pmatrix}\infty\ 0\ \infty\ \infty\ \infty\ \infty\end{pmatrix} \rightarrow \begin{pmatrix}\infty\ \infty\ \infty\ \infty\ \infty\ \infty\end{pmatrix}$$

$$\pi = \begin{pmatrix}0\ 0\ 0\ 1\ 1\ 4\end{pmatrix} \rightarrow \begin{pmatrix}0\ 0\ 4\ 1\ 1\ 4\end{pmatrix} \rightarrow \begin{pmatrix}0\ 1\ 4\ 1\ 1\ 4\end{pmatrix}$$

$$\mathbf{d} = \begin{pmatrix}0\ 6\ 5\ 0\ 0\ 0\end{pmatrix} \rightarrow \begin{pmatrix}0\ 6\ 0\ 0\ 0\ 0\end{pmatrix} \rightarrow \begin{pmatrix}0\ 0\ 0\ 0\ 0\ 0\end{pmatrix}$$

6.3.1.2 Python Implementation

The Python implementation of the algebraic algorithm is shown below with few modifications. The current distance of nodes to the source node is kept in vector d which is initialized to all -1s, parents of nodes are stored in dictionary *parents* and the variable *count* is initialized to $n - 1$ and is decremented each time a node is included in the MST. The next node in MST is kept in vector *nextv* which is initialized to all -1s. The procedure starts by first finding the endpoint of MWOE from the source vertex f input to the procedure and *nextv* is updated by this vertex in line 17, the source vertex is stored in s which is a list of visited vertices. The next vertex to be added to the *weight* variable stores the current total weight of MST edges discovered. Each iteration of the *while* loop finds the index of the minimum weight node first, adds its weight to the *weight* variable. The *while* loop continues until *count* becomes 0. Each time an endpoint $y1$ of a MWOE is found in the adjacency matrix A, a check is made to determine whether $y1$ is already in the MST in 33–36 and the rest of the loop is skipped if this is the case.

```
1   ################################################################
2   #                  Jasnik-Prim's Algorithm                     #
3   ################################################################
4
5   import numpy as np
6   import math as math
7
8   def Prim(A,f):
9     n=len(A)
10    T = A.copy()
11    parents = {}      # initialize
12    d = np.full(n,inf)
13    weight = 0
14    parents[f] = 0
15    d[f] = min(A[f,:])   # find first MWOE
16    nextv = np.full(n,-1)
17    nextv[f] = np.argmin(A[f,:])
18    count = n -1
19    s = []
20    s.append(f)
21
22    while count > 0:    # main loop
23        u = np.argmin(d)
24        weight = weight + d[u]
25        v = nextv[u]
26        parents[v] = u
27        s.append(v)
28
29        A[u,v] = A[v,u] = inf  # update
30        for j in range (0,n):   # check for a loop
31            m1 = min(A[u,:])
32            y1 = np.argmin(A[u,:])
33            if y1 in s:
34                A[u,y1] = inf
```

```
35              A[y1,u] = inf
36              continue
37          weight = weight + d[u]
38          d[u] = m1
39          nextv[u] = y1
40          for j in range (0,n):
41              if (T[v,j] != inf) and (j in s) and (nextv[j]==v):
42                  A[v,j] = inf
43                  A[j,v] = inf
44                  d[j] = min(A[j,:])
45                  nextv[j] = np.argmin(A[j,:])
46              m2 = min(A[v,:])
47              y2 = np.argmin(A[v,:])
48              if y2 in s:
49                  A[v,y2] = inf
50                  A[y2,v] = inf
51                  continue
52          d[v] = m2
53          nextv[v] = y2
54          count = count -1
55      return weight, parents
56
57  if __name__ == '__main__':
58      inf = math.inf
59      B = np.array([[inf,8,inf,inf,2,inf,inf,inf,inf,inf,inf],
60                    [8,inf,7,3,1,inf,inf,inf,14,inf,inf],
61                    [inf,7,inf,5,inf,6,4,inf,inf,inf,inf],
62                    [inf,3,5,inf,6,inf,13,9,inf,inf,inf],
63                    [2,1,inf,6,inf,inf,inf,inf,inf,inf,inf],
64                    [inf,inf,6,inf,inf,inf,10,inf,inf,inf,inf],
65                    [inf,inf,4,13,inf,10,inf,inf,inf,inf,11],
66                    [inf,inf,inf,9,inf,inf,inf,inf,inf,inf,inf],
67                    [inf,14,inf,inf,inf,inf,inf,inf,inf,12,inf],
68                    [inf,inf,inf,inf,inf,inf,inf,inf,12,inf,inf],
69                    [inf,inf,inf,inf,inf,inf,11,inf,inf,inf,inf]])
70
71      Weight, Parents = Prim(B,0)
72      print ("Total Weight:", Weight)
73      print ("Parents:", Parents)
74  >>>
75  Total Weight: 67.0
76  Parents: {0: 0, 4: 0, 1: 4, 3: 1, 2: 3, 6: 2, 5: 2, 7: 3, 10: 6,
77  8: 1, 9: 8}
```

Running of this algorithm from vertex 0 in the graph of Fig. 6.4 resulted in the MST edges shown in bold pointing to parents as output by the algorithm in the last lines. This MST with a total weight of 67 is unique as edge weights are distinct. There are two main loops in the algorithm with the *while* loop running $O(m)$ times and the *for* loop in lines 41–52 running $O(n)$ times totalling $O(mn)$ times for this implementation excluding the time taken to find the minimum value in a row of

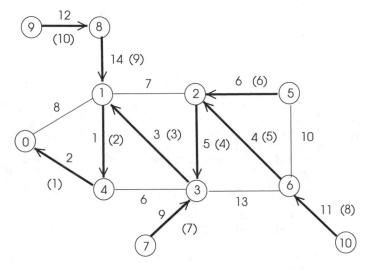

Fig. 6.4 Running of JP Algorithm in a sample graph

adjacency matrix. For a large array, finding the minimum value and then finding the argument of this minimum value can be performed in parallel.

6.3.1.3 Parallel Implementation

For a large graph, parallel processing will improve performance. The MST algorithms including JK algorithm do not have matrix multiplication as the core operation. Thus we need to search other operations in these algorithms for parallel processing. As a first attempt, we can see finding the minimum value and argument of the minimum value of matrix A matrix may be done in parallel. We will sketch parallel algorithms in shared memory and distributed memory models in the following sections. For various other problems throughout the text, matrix multiplication is a fundamental operation and we will frequently state that at least a partial parallel version of the sequential algorithm may be obtained simply by parallelizing matrix multiplication.

Shared Memory Model

We will do column-wise partitioning of the row u of matrix A to k threads as shown in Fig. 6.5. The row u is searched concurrently by all threads. We implement the supervisor/worker thread model in shared memory implementation with synchronization provided by operating system structures such as semaphores. Let the number of created threads be k, then we will have $\lceil n/k \rceil$ column partition assigned to each thread based on its identity. Each thread T_i searches the columns assigned to it for row u under consideration and finds the minimum distance value in its assigned columns

Fig. 6.5 Partitioning of distance matrix

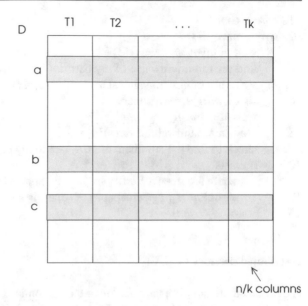

for this row. It then sends this value to the supervisor thread which calculates the minimum of these minimum values, sends it to all threads which attempt to find the argument of this minimum value in their column partitions. The supervisor then assigns a new row to be searched in next iteration and sends this value to all threads which continue the search until all rows (vertices) are served.

The supervisor thread performs the following algorithm steps in synchronous rounds. A new vertex is added to MST at the end of each round until all vertices are visited. There are two arrays of semaphores, $min\_vals[k]$ and $min\_args[k]$ for k threads to wait for column partitions and to wait for minimum value found, with each element belonging to a single thread.

1. **while** $count > n - 1$
2. $next \leftarrow u$
3.   **signal** min_sems semaphores of threads
4.   **wait** at my min_sem for completion
5.   **find** minimum of minimum values from $min\_values[k]$
6.   **write** this value to global memory min_val
7.   **signal** min_args semaphores of threads
8.   **wait** at my arg_sem
9.   **update** d and A
10. $count \leftarrow count - 1$

A worker thread performs the following algorithm steps at each round under the control of the supervisor:

1. *count* ← *n* − 1
2. **while** *count* > 0
3.   **wait** at my *min_sems*[*me*]
4.   **find** the minimum value of my partition of row
5.   **write** my minimum value at *min_values*[*me*]
6.   **signal** *min_sem* of supervisor
7.   **wait** at my *min_args*[*me*]
8.   **read** minimum value from *min_val*
9.   **check** to find the argument of *min_val* in my columns
10.   **if** exists
11.   **write** argument of *min_val* in min_arg
12.   **signal** *min_arg* semaphore of supervisor
13.   *count* ← *count* − 1

Distributed memory

A distributed memory implementation of the algebraic JP algorithm may be realized again by sending partial columns of all rows to worker processes, this time using messages and supervisor/worker model. Note that we do not need to synchronize the workers with the supervisor since messages provide this synchronization. A possible algorithm steps for the supervisor is as follows.

1. **while** *count* > *n* − 1
2.   *next* ← *u*
3.   **send** column partitions to workers
4.   **receive** minimum values
5.   **find** minimum of minimum values *min_val*
6.   **send** *min_val* to workers
7.   **receive** a single argument
8.   **update** *d* and *A*
9.   *count* ← *count* − 1

Waiting and signalling of semaphores are replaced by *send* and *receive* routines. A worker process performs the following in synchronization with the supervisor.

1. *count* ← *n* − 1
2. **while** *count* > 0
3.   **receive** my column partition
4.   **find** the minimum value of my partition
5.   **send** my minimum value to supervisor
6.   **receive** global minimum value *min_val*
7.   **check** to find the argument of *min_val* in my columns
8.   **if** exists
9.   **send** argument of *min_val* to supervisor

10.   *count* ← *count* − 1

6.3.2   Kruskal's Algorithm as a Matroid

Kruskal's MST algorithm is simply the matroid described in Sect. 4.7.3, it is a greedy algorithm that sorts the edges of a graph with respect to their weights with the lightest edge first [1]. It then sets a set S to be empty and for each edge e in the sorted list, includes e in T as long as e does not create a cycle with the existing edges in T as in the below algorithm steps.

1. **Input**: An undirected weighted graph $G = (V, E, w)$
2. **Output**: An MST $T = (V, E')$ of G
3. $T \leftarrow \emptyset$
4. Sort edges of G in non-increasing order and place them in a queue Q.
5. **repeat**
6.   Remove the first edge (u, v) from Q and add it to T if it does not form a cycle with the edges edge that are already included in T.
7. **until** there are $n - 1$ edges in T.

We can use union-find data structure to find whether the two endpoints of the selected edge are in the same set (the current MST fragment). It can be shown that this algorithm is correct and has a time complexity of $O(m \log n)$ [1].

Python Implementation

The difficult part of Kruskal's algorithm is detecting a cycle when an edge is to be added to the MST. Another point to consider is the growing of MST in different regions of the graph which we will call *components* of the MST. In the Python implementation shown below, we have the *comps* list which shows the component a vertex belongs. This list is initialized to all −1s and if both endpoints of an edge have −1 value in the comps list, it does not belong to any component.

The algorithm starts by sorting all edges into the queue e and each time an edge (u, v) is dequeued from e, a check is made to classify the edge as follows. The main *while* loop runs until queue e becomes empty.

- If both u and v have the same component value of −1, (u, v) is the initiator of a new component and the component value is set to the minimum of the two vertex values in lines 25–26.
- If both u and v have the same component value other than −1, the edge (u, v) will form a cycle within the component, thus, (u, v) is discarded and we move to the next iteration in lines 27–28.

- If one of the vertices u and v has a component value -1 and the other one has a valid value showing it belongs to an existing component, we merge (u, v) to the existing component in lines 29–32.
- The last possible case is when edge (u, v) is between two components in which case we need to merge the components. This condition is tested in line 33 and components are merged by setting the component value of u and v to the smaller one of the merged components.

```python
##############################################################
#                    Kruskal's Algorithm                     #
##############################################################

import numpy as np
import math as math

def Kruskal(D):

   n=len(D)
   e = []                  # edge list
   c = [-1]*n              # components list

   for i in range(0,n):    # initialize edge list
       for j in range (i+1,n):
           if D[i,j] != inf:
               e.append(tuple((i,j,D[i,j])))
   e.sort(key=lambda x:x[2]) # edges are sorted
   s = []   # edges in MST, i, j and w(i,j)
   while len(e) > 0:    # iterate until list is empty
       new = e.pop(0)
       x = new[0] # find endpoints
       y = new[1]
       if c[x] == c[y] == -1: # new component
           c[x] = c[y] = min(x,y)
       elif c[x] == c[y]:    # cycle detected
           continue
       elif c[x] == -1 and c[y] != -1: # merge edge
           c[x] = c[y]
       elif c[x] != -1 and c[y] == -1: # merge edge
           c[y] = c[x]
       elif c[x] != -1 and c[y] != -1: # merge components
           m = min(c[x],c[y])
           if c[x] > m:
               for i in range(0,n):
                   if  D[x,i] != inf and c[i] !=-1:
                       c[i] = m
           if c[y] > m:
               for i in range(0,n):
                   if  D[y,i] != inf and c[i] !=-1:
                       c[i] = m
           c[y] = c[x] = m
```

```
43        s.append(new)
44     return s
45
46  if __name__ == '__main__':
47     inf = math.inf
48     B = np.array([[0,10,inf,inf,inf,inf,4,inf,inf],
49                    [10,0,1,inf,inf,5,3,inf,inf],
50                    [inf,1,0,6,inf,13,inf,inf,inf],
51                    [inf,inf,6,0,11,8,inf,9,12],
52                    [inf,inf,inf,11,0,2,inf,inf,inf],
53                    [inf,5,13,8,2,0,7,inf,inf],
54                    [4,3,inf,inf,inf,7,0,inf,inf],
55                    [inf,inf,inf,9,inf,inf,inf,0,inf],
56                    [inf,inf,inf,12,inf,inf,inf,inf,0]])
57     edges = Kruskal(B)
58     print ("Kruskal MST:", edges)
59  >>>
60  Kruskal MST: [(1, 2, 1.0), (4, 5, 2.0), (1, 6, 3.0), (0, 6, 4.0),
61  (1, 5, 5.0), (2, 3, 6.0), (3, 7, 9.0), (3, 8, 12.0)]
```

Running of this algorithm in the simple weighted graph of Fig. 6.6 results in the MST shown in bold lines as output of the algorithm in the form of (u, v, w) where (u, v) is the edge and w is the weight of (u, v). The time spent is dominated by the initialization which is $O(n^2)$ to form edge list in this implementation.

6.3.3   Boruvka's Algorithm

Boruvka's algorithm builds an MST of a weighted graph using a greedy approach by adding the lightest edges incident on components. Initially, each vertex of the graph is a component. This algorithm consist of two main steps; finding minimum weight edges out of the components and contracting components. The lightest outgoing edge from a component is part of the MST as in Prim's algorithm. Specifically, the following steps are performed.

Fig. 6.6 A sample graph to test Kruskal's algorithm

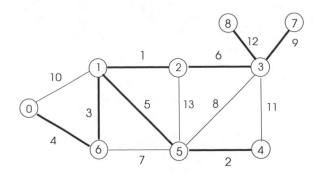

1. **Input**: An undirected graph $G = (V, E)$
2. **Output**: An MST $T = (V, E')$ of G
3. $T \leftarrow \emptyset$
4. Each vertex $v \in V$ is a component initially.
5. **repeat**
6.   find the lightest (u, v) edge that is incident on a component
7. $T \leftarrow T \cup \{(u, v)\}$
8.   Direct arrows from lightest edges to the components.
9.   Contract a component with the lightest edge that points to it iteratively.
10. **until** there is only one component

Note that contraction at step 9 may result in contraction of two or more components in chain. Also, if the lightest edge between two components is the same, these components are merged. Running Boruvka's algorithm in the graph of Fig. 6.3 is depicted in Fig. 6.7. Initially each vertex is a component and a vertex u forms a component with a vertex v that is the endpoint of the lightest edge (u, v) incident to vertex u and such lightest edges are included in the MST. We can see that the two components formed this way are $C_1 = \{6, 4, 3\}$ and $C_2 = \{1, 2, 5\}$ shown in (a), and the lightest edge between this components is (1,4) which is added to the current MST in (b) to form the final MST which is the same MST as in Fig. 6.3 since edge weights are distinct.

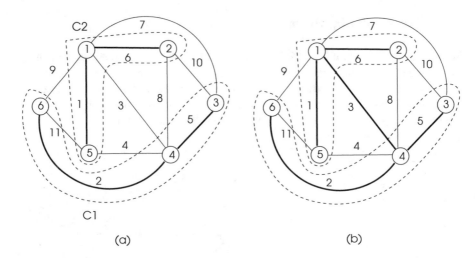

(a)   (b)

Fig. 6.7 Implementation of Boruvka's algorithm in the graph of Fig. 6.3

Sketch of A Parallel Algorithm

Boruvka's algorithm allows independent operations in different regions of the graph under consideration, thus, it has potential to be parallelized. Considering its operation, we can see that finding lightest edges incident to components and contraction of the components are the steps that can be performed in parallel. The weighted adjacency matrix D for the graph of Fig. 6.7 is shown below. Two components formed after the first iteration with their outgoing edges are shown next to this matrix and we find the edge (1,4) with weight 3 is the lightest for both which can be used to contract the components into a single one which terminates the algorithm.

$$
D = \begin{array}{c} \\ 1 \\ 2 \\ 3 \\ 4 \\ 5 \\ 6 \end{array}
\begin{array}{c} 1\ \ 2\ \ 3\ \ 4\ \ 5\ \ 6 \\
\begin{pmatrix}
0 & 6 & 7 & 3 & 1 & 9 \\
6 & 0 & 10 & 8 & 0 & 0 \\
7 & 10 & 0 & 5 & 0 & 0 \\
3 & 8 & 5 & 0 & 4 & 2 \\
1 & 0 & 0 & 4 & 0 & 11 \\
9 & 0 & 0 & 2 & 11 & 0
\end{pmatrix}
\end{array}
\rightarrow
\begin{array}{c} \\ C1 \\ C2 \end{array}
\begin{array}{c} C1\ \ \ \ \ \ C2 \\
\begin{pmatrix}
3,4,8,9 & 0 \\
0 & 3,4,8,9
\end{pmatrix}
\end{array}
\rightarrow C
$$

Graph contraction can be performed by sparse matrix by matrix multiplication. The sequential procedure for this purpose is shown in Algorithm 6.4 [2]. The sparse matrix multiplication can be performed in parallel.

Algorithm 6.4 *Spanning Tree Algorithm*

1: **procedure** CONTRACT(G,labels)
2:     $n = length(G)$
3:     $m = max(labels)$
4:     $S = sparse(labels, 1 : n, 1, m, n)$
5:     $C = S * G * S'$
6:     return (C)
7: **end procedure**

6.4 Chapter Notes

The main graph problems we reviewed in this chapter are forming a general spanning tree of an unweighted connected graph and building an MST of a weighted connected graph in this chapter. Another spanning tree algorithm makes use of the tree property of a tree as follow. The spanning tree T is set to all edges of the graph G initially and edges are removed one by one until any edge removal leaves T disconnected. However, checking whether G is disconnected or not is not a trivial task. A different spanning tree algorithm initiates spanning trees around nodes initially and merges

Fig. 6.8 A sample graph for
Exercise 2

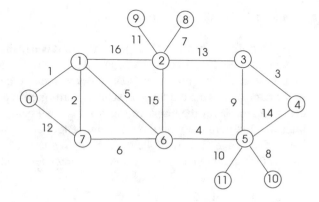

Fig. 6.9 A sample graph for
Exercise 1

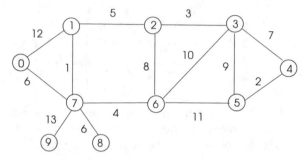

them iteratively [1]. Parallel formation of the spanning tree using the adjacency
matrix is costly, however, fast parallelization is possible as shown for shared memory
C/POSIX implementation and distributed memory parallel implementation. We tried
to be as general as possible when forming the parallel algorithms, however, the
C/POSIX code is directly implementable.

Matroid theory is claimed to be as general as graph theory and concepts introduced
in weighted matroids can be used to construct MSTs when the set under consideration
is the edge set of a graph with weights. In fact, Kruskal's MST algorithm is basically
a matroid implementation in a graph. We described ways to parallelize Jarnik-Prim
and Boruvka's algorithms. Data structures play an important role when designing
sequential algorithms for these applications and in any general application. Parallel
implementations require careful assignments of matrix data and synchronization
points.

Exercises with Programming

1. Show the iteration steps of Prim's algorithm in the graph of Fig. 6.8.
2. Provide a shared memory parallel JP algorithm using the Python *multiprocessing*
   module using the method described in Sect. 6.3.1.3.

3. Provide a distributed memory parallel JP algorithm using the Python *mpi4py* module using the method described in Sect. 6.3.1.3.
4. Show the iterations of Kruskal's MST algorithm in the graph of Fig. 6.3.
5. Propose a method to parallelize Kruskal's algorithm.
6. Provide the Python code for sequential Boruvka's algorithm using a similar approach to merge components as in the Python program for Kruskal's algorithm.
7. Show the iteration steps of Boruvka's algorithm in the graph of Fig. 6.9.
8. Write a Python function to perform edge contraction as part of Boruvka's algorithm using sparse matrix multiplication from *numpy* library.
9. Provide the pseudocode of a method to parallelize Boruvka's algorithm.

References

1. K. Erciyes, *Guide to Graph Algorithms: Sequential, Parallel and Distributed* (Springer, 2018)
2. J. Kepner, J. Gilbert, *Graph Algorithms in the Language of Linear Algebra* (2014)

Shortest Paths

7

Abstract

Finding the shortest shortest path between two vertices in a weighted graph has many applications. A fundamental implementation involves sending packets over shortest routes, therefore, in shortest possible time between two routers in a computer network. This so-called *routing* problem is commonly solved by converting a graph routing algorithm to a distributed one in the form of a network protocol to be executed by each router. We look at basic graph algorithms to find shortest paths in a weighted graph in this chapter.

7.1 Introduction

An edge-weighted or simply a weighted graph has weights associated with its edges. Finding shortest paths between two vertices in a weighted graph has many applications such as finding the least cost route between two routers in a computer network. In fact, this so called routing is a fundamental problem in such networks. In the single source shortest path (SSSP) version of this problem, we need to find distances from a source node to all other nodes in a weighted graph. The all-pairs-shortest-paths (APSP) deals with finding distances among all vertex pairs in the network graph.

A shortest path tree with the root as the source node is formed at the end of the SSSP algorithm with the sum of the weights along the path from the source vertex s to a vertex v is minimum among all paths from s to v. We will review two basic algorithms; Dijkstra's algorithm and Bellman-Ford algorithm with their algebraic and Python versions to solve SSSP problem in a weighted graph in this chapter. The APSP problem can be solved by Floyd-Warshall algorithm that finds shortest paths between every vertex pair. Both Bellman-Ford and Floyd-Warshall algorithms use dynamic programming method by storing the intermediate best-known routes and enhancing them at each iteration to find the best route in the end. These algorithms use a technique called *relaxation* to improve the routes found as shown in the following

© Springer Nature Switzerland AG 2021

K. Erciyes, *Algebraic Graph Algorithms*, Undergraduate Topics in Computer Science, https://doi.org/10.1007/978-3-030-87886-3_7

code with $d(u)$ as the distance of the node u to the source, $N(u)$ is the neighbor set of u and $p(u)$ is the parent of vertex u.

1. **for all** $v \in N(u)$
2.   **if** $d(u) > d(v) + w(u, v)$
3. $du) = d(v) + w(u, v)$
4. $p(u) = v$

The main idea is checking whether a shorter path than the previously computed one exists from a neighbor vertex v of u, then the shortest path is formed through that vertex and the distance of u is updated to be the sum of the distance of v plus the weight of the edge joining u and v.

For shortest paths in an unweighted undirected or directed graph, the breadth-first-search (BFS) algorithm may be used which finds the distances of vertices to a source node as the minimum number of edges between a vertex and the source. We start this chapter with the BFS algorithm and then describe SSSP and APSP algorithms for weighted graphs.

7.2 Breadth-First-Search

The BFS algorithm is executed from a source vertex s and first visits all neighbors of s labeling them with level 1, then it visits neighbors of these neighbors which are two hops away from s, labeling them with level 2 and it continues in this manner until all vertices are visited.

7.2.1 The Classical Algorithm

The classical BFS algorithm is executed from a source vertex s and finds the levels of vertices to s and the parent of each vertex on the BFS tree. The algorithm depicted in Algorithm 7.1 inputs an unweighted directed or undirected graph G, initializes all vertex distances to s as infinity and parent of each vertex is set to -1 with parent of s set to itself and its distance to itself is made zero. Starting from vertex s, the neighbors of a visited vertex u are stored in a queue Q and these vertices are taken from Q one by one, if a neighbor vertex v of vertex u is not visited, its distance is set to distance of u plus 1 and its parent is set to vertex u. This process continues until all vertices are visited which is verified by an empty queue. Note that checking whether a vertex is visited or not can be done conveniently by checking its current distance with distance infinity meaning it is not visited.

The running of this algorithm in a sample graph is shown in Fig. 7.1 with the formed tree edges at each iteration. The BFS tree structure may vary depending on the order of dequeuing from Q; for example, if vertex 1 is dequeued first from Q in (b), then we would have (1,2) as the BFS tree edge instead of (6,2) edge. The time complexity of this classical BFS algorithm is $O(n)$ for the initialization part and $O(m)$ for testing each edge in both directions for a total of $O(n + m)$ time [1].

Algorithm 7.1 *BFS*

1: **Input** : $G = (V, E)$ ▷ connected, Uweighted graph G and a source vertex s
2: **Output** : $l[n]$ and $p[n]$ ▷ levels and parents of vertices in shortest path tree
3: **for all** $v \in V \setminus \{s\}$ **do**   ▷ initialize all vertices except source s
4:     $d[v] \leftarrow \infty$
5:     $p[v] \leftarrow -1$
6: **end for**
7: $d[s] = 0; p[s] = s$ ▷ initialize source
8: $Q \leftarrow s$
9: **while** $Q \neq \emptyset$ **do**
10:     $u \leftarrow deque(Q)$
11:     **for all** $(u, v) \in E$ **do**   ▷ update neighbor levels
12:         **if** $l[v] = \infty$ **then**
13:             $l[v] \leftarrow l[u] + 1$
14:             $p[v] \leftarrow u$ ▷ update parent
15:             $enque(Q,v)$
16:         **end if**
17:     **end for**
18:     $S \leftarrow S \setminus \{u\}$ ▷ remove u from searched
19: **end while**

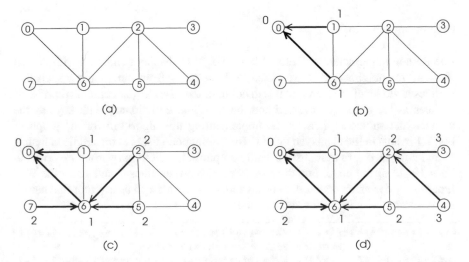

Fig. 7.1 Running of BFS algorithm on a sample graph

7.2.2   The Algebraic Algorithm

The algebraic BFS algorithm uses the adjacency matrix A of a graph $G = (V, E)$. We first initialize a vector x which consists of all zeros except the index of the source vertex s that we want to run the algorithm from; we then form the matrix $A' = A + I$ and then $A^T \cdot x$ selects all nodes which are at distance 1 (level 1) from the source vertex. Multiplying the vector x with the matrix A^2 gives the vertices that are at most 2 hops. In general, the product $A^k \cdot x$ will give neighbors that are at most k hops away and the multiplication should be performed boolean as in Algorithm 7.2.

Algorithm 7.2 *Algebraic BFS*

1: **Input** : Adjacency matrix $A_{n,n}$ of a graph $G = (V, E)$ ▷ connected, unweighted graph G and a
    source vertex s
2: **Output** : N, *visited*, *levels*   ▷ a matrix that shows level i vertices at its column i, the visited
    vertices in sequence and their levels
3: $x[n] \leftarrow 0$
4: $x[s] \leftarrow 1$
5: $A \leftarrow A + I$
6: **for** $i = 1$ to n **do**
7:    $N \leftarrow A^T \cdot x$
8:    *levels*$[i] \leftarrow$ column i of N
9:    *visited* $\leftarrow$ *visited*$\cup$ {visited nodes at this iteration}
10: **end for**

Python Implementation

The Python implementation below has matrix N to display neighbors with $N[: k]$ showing the neighbors that are exactly k hops away from the source s such that vertices at ith level are stored at its ith column, the *visited* list contains the visited vertices as the graph is traversed and the *levels* is a dictionary with keys as the vertices and values as their levels. Implementing this algorithm for the graph of Fig. 7.1 results in the displayed output. The output node list is extracted from matrix N and converted into more user friendly output in the dictionary structure *levels* in lines 19–23 of the code. Initialization is done between lines 7 and 13 in the BFS function by declaring matrix N to be all zeros and vector x to be a vector that has all zeros except the index of the source vertex.

```
1   ####################################################################
2   #                   Breadth-First-Search Algorithm                 #
3   ####################################################################
4   import numpy as np
5
6   def BFS(A,s):
```

```
7    n = len(A)
8    N = np.zeros((n,n), dtype=int)  #  neighbor matrix N
9    visited = []     # shows all visited notes without levels
10   levels = {}      # shows nodes at each level
11   x = np.zeros((n), dtype=bool).T
12   x[s] = True
13   A_T = A.transpose()
14   y_old = x
15   N[:,0] = (x.astype(int)).T     # column 0 is the source
16   levels[0] = [s]      # 0 level has the source
17   visited.append(s)    # source is marked
18   for i in range(1,n):
19       y_new = np.dot(A_T,x)    # neighbors i hop away
20       t = y_new.astype(int) - y_old.astype(int)
21       N[:,i] = t.T                # save in 'i'th column of N
22       x = y_new                   # update
23       y_old = y_new
24       nodes = np.argwhere(N[:,i]==1) # get level nodes
25       if np.all((nodes==0)): # stop if all nodes are visited
26           break
27       listed = [i[0] for i in nodes.tolist()] #save in list
28       levels[i] = listed
29       visited.extend(listed)
30   return N, visited, levels
31
32  if __name__ == '__main__':
33      B = np.array([[1,1,0,0,0,0,1,0], [1,1,1,0,0,0,1,0],
34          [0,1,1,1,1,1,1,0], [0,0,1,1,0,0,0,0],
35          [0,0,1,0,1,1,0,0], [0,0,1,0,1,1,1,0],
36          [1,1,1,0,0,1,1,1], [0,0,0,0,0,0,1,1]], dtype=bool)
37      neighbors, visited, levels = BFS(B,0)
38      print ("Neighbors = ")
39      print (neighbors)
40      print ("Visited = ", visited)
41      print ("Levels = ", levels)
42
43  >>>
44  Neighbors =
45  [[1 0 0 0 0 0 0 0]
46   [0 1 0 0 0 0 0 0]
47   [0 0 1 0 0 0 0 0]
48   [0 0 0 1 0 0 0 0]
49   [0 0 0 1 0 0 0 0]
50   [0 0 1 0 0 0 0 0]
51   [0 1 0 0 0 0 0 0]
52   [0 0 1 0 0 0 0 0]]
53  Visited =   [0, 1, 6, 2, 5, 7, 3, 4]
54  Levels =   {0: [0], 1: [1, 6], 2: [2, 5, 7], 3: [3, 4]}
```

The initialization of matrix N and taking transpose of matrix A both take $O(n^2)$ operations which are the dominant times at initialization. The main for loop in lines 18–30 is executed $O(n)$ times and this loop is exited at line 26 if all neighbors of the source vertex are visited. The matrix-vector multiplication at line 19 takes $O(n^2)$ time; thus, we have $O(n^3)$ time for this algorithm. However, parallelization of matrix multiplication is possible with row-wise or column-wise partitioning methods described in Chap. 5.

7.3   Dijkstra's Algorithm

Dijkstra's algorithm finds shortest paths from a given source to all other nodes in a weighted graph with non-negative weights. Starting from the source vertex, the vertex with the lowest weight is added to the searched vertices. Whenever a vertex u is decided to be included in the shortest path, its neighbors are tested. If a neighbor v of u has a lower distance to source s over u, its distance is updated and its parent is set to vertex u using relaxation. The pseudocode of this algorithms is shown in Algorithm 7.3 where vector d holds the current best known distances of vertices to the source vertex and p is a vector showing current parents. The algorithm finishes when distances of all vertices are determined.

Running of this algorithm from the source vertex 0 is depicted in Fig. 7.2. Each iteration results in a new discovered edge shown in emphasized line that belongs to the shortest path tree from the source.

The time complexity of the algorithm can be improved by using priority queues to $O(nlogn + m)$.

Algebraic Algorithm with Python

The algebraic implementation of Dijkstra's algorithm keeps the distance values in dictionary d which is initialized to all 99 values for infinity and parents are kept in dictionary p which is initialized to all -1 values except the source vertex s. These two dictionaries are returned to the main program which outputs them as displayed. We use an auxiliary vector dd to identify the visited vertices; every time a vertex is assigned to be on the shortest path to the source, we change its distance in dd to infinity (line 21) to exclude it from further checking. We use the distance matrix of the graph of Fig. 7.2 as the input to this algorithm and the output is consistent with the obtained tree of Fig. 7.2.

Algorithm 7.3 *Dijkstra's Algorithm*

1: **Input** : $G = (V, E, w)$ ▷ connected, weighted graph G and a source vertex s
2: **Output** : $d[n]$ and $p[n]$ ▷ distances and parents of vertices in shortest path tree
3: **for all** $v \in V \setminus \{s\}$ **do**   ▷ initialize all vertices except source s
4:     $d[v] \leftarrow \infty$
5:     $p[v] \leftarrow -1$
6: **end for**
7: $d[s] = 0; p[s] = s$ ▷ initialize source
8: $S \leftarrow V$
9: **while** $S \neq \emptyset$ **do**
10:     **find** $u \in S$ with minimum distance value
11:         **for all** $(u, v) \in E$ **do**   ▷ update neighbor distances to v
12:             **if** $d[v] > d[u] + w(u, v)$ **then**
13:                 $d[v] \leftarrow d[v] > d[u] + w(u, v)$
14:                 $p[v] \leftarrow u$ ▷ update parent
15:             **end if**
16:         **end for**
17:     $S \leftarrow S \setminus \{u\}$ ▷ remove u from searched
18: **end while**

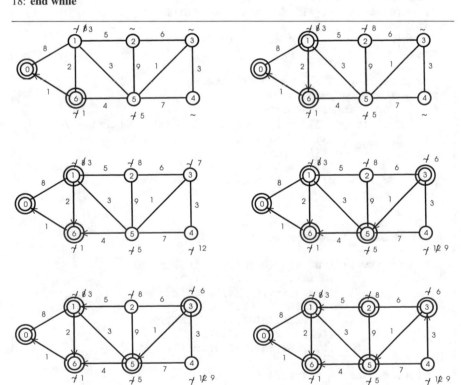

Fig. 7.2 A sample graph

```
###############################################################
#                        Dijkstra's Algorithm                 #
###############################################################

import numpy as np
import math
inf = math.inf

def Dijkstra(D,s):

    n=len(D)
    d = {}

    # initialize distances
    [d.setdefault(x,[inf]) for x in range(n)]
    d[s] = 0
    p = {}
    # initialize parents
    [p.setdefault(x,[-1]) for x in range(n)]
    p[s] = s
    dd = [inf]*n          # auxiliary distance vector
    dd[s] = 0

    for x in range(0,n):  # do n-1 steps
        i = np.argmin(dd)
        for j in range(0,n):
            if D[i,j] != inf and d[j] > d[i] + D[i,j]:
                d[j] = d[i] + D[i,j]
                dd[j] = d[j]
                p[j] = i
        dd[i] = inf        # remove i from search
    return d, p


if __name__ == '__main__':

    B = np.array([[inf,8,math.inf,inf,inf,inf,1],
                  [8,inf,5,inf,inf,3,2],
                  [inf,5,inf,6,inf,9,inf],
                  [inf,inf,6,inf,3,1,inf],
                  [inf,inf,inf,3,inf,7,inf],
                  [inf,3,9,1,7,inf,4],
                  [1,2,inf,inf,inf,4,inf]])

    Dist, Parents = Dijkstra(B,0)
    print ("Distances:", Dist)
    print ("Parents:", Parents)
>>>
Distances: {0: 0, 1: 3.0, 2: 8.0, 3: 6.0, 4: 9.0, 5: 5.0, 6: 1.0}
Parents: {0: 0, 1: 6, 2: 1, 3: 5, 4: 3, 5: 6, 6: 0}
```

Each iteration of the algorithm assigns the shortest path of a new vertex to the source vertex s and thus, it suffices to have *for* loop executing $n - 1$ times excluding the source. The time complexity of this implementation is $O(n^2)$ due to two nested *for* loops in lines 24–31.

7.4   Bellman-Ford Algorithm

Bellman-Ford algorithm finds shortest paths from a source vertex s to all other vertices in a connected weighted and undirected or directed graph. It uses dynamic programming principle by storing the intermediate results and using these computed distance values to reach the optimal distance values to the source. This algorithm may be used in graphs with negative weights and can detect if a negative cycle exists in the graph under consideration.

The classical Bellman-Ford algorithm uses *relaxation* to assign distances to vertices and assigning their parents along the paths to the source. Each edge is tested in each iteration whether its distance to the source can be changed to a lesser value by the previously computed distance values as shown in Algorithm 7.4. Time complexity of the algorithm in this straightforward implementation is $O(nm)$ since each edge is inspected for m times in each iteration of the outer *for* loop which runs $n - 1$ times.

Running of this algorithm in a small graph is depicted in Fig. 7.3 with the shortest path tree from source vertex 0 shown in bold.

Python Implementation

The Python implementation is similar in structure to Algorithm 7.4. We have two nested *for* loops in lines 18 and 20 and vector addition in line 21 takes n steps. Finding the minimum value and finding the index of this minimum value take $O(n)$ time inside the inner loop. Thus, total time complexity for this Python implementation is $O(n^3)$. Comparing with the classical algorithm complexity, we find $O(nm)$ can be interpreted as $O(n^3)$ for a dense graph since the number of edges m in a dense graph is $O(n^2)$, thus, Python implementation has a comparable time complexity. The output obtained is the same as in Fig. 7.3.

Algorithm 7.4 *Bellman_Ford*

```
 1: procedure BELLMAN_FORD(G, s)
 2:     for all {v} ∈ V do                                    ▷ initialize
 3:         d[v] ← ∞
 4:         p[v] ← Ø
 5:     end for
 6:     d[s] ← 0                                              ▷ s has 0 cost
 7:     for k = 1 to n − 1 do
 8:         for all (u, v) ∈ E do                             ▷ relaxation
 9:             if d[v] > d[u] + w[u, v] then
10:                 d[v] = d[u] + w[u, v]
11:                 p[v] ← u
12:             end if
13:         end for
14:     end for
15:     for all (u, v) ∈ E do
16:         if d[v] > d[u] + W[u, v] then
17:             return Ø                                      ▷ a negative cycle detected
18:         end if
19:     end for
20:     return d, P
21: end procedure
```

```python
1   ###############################################################
2   #                 Bellman-Ford Algorithm                      #
3   ###############################################################
4
5   import numpy as np
6   import math
7   inf = math.inf
8
9   def Bellman_Ford(A,s):
10
11      n=len(A)
12      d = np.full(n,inf)      # initialize distance vector
13      parents = {}            # holds parents
14      parents[0] = s
15      d[s] = 0
16      d_old = d.copy()
17
18      for k in range(1,n):            # do n-1 steps
19          print ("vector d:", d)
20          for j in range(0,n):
21              sum_d = d_old + A[:,j].T # add for node j
22              min_val = min(sum_d)
23              if min_val < d[j]:       # update d
24                  d[j] = min_val
25                  parents[j] = np.argmin(sum_d)
```

```
26        d_old = d.copy()
27     return d, parents
28
29
30 if __name__ == '__main__':
31
32     B = np.array([[inf,8,inf,inf,1],
33                   [8,inf,7,3,2],
34                   [inf,7,inf,1,inf],
35                   [inf,3,1,inf,6],
36                   [1,2,inf,6,inf]])
37
38     D, Parents = Bellman_Ford(B,0)
39     print ("Distances:", D)
```

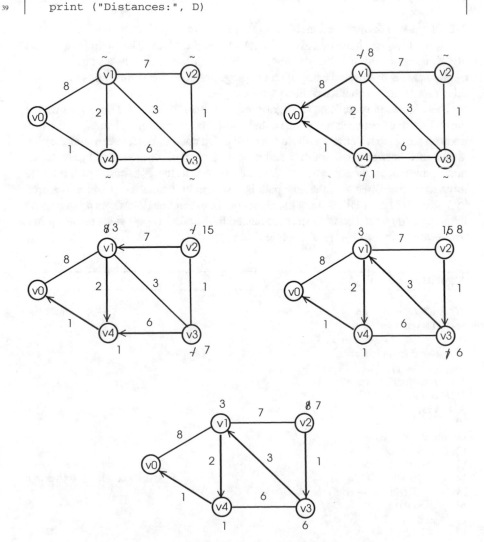

Fig. 7.3 Running of Bellman-Ford algorithm in a sample graph

```
40    print ("Parents:", Parents)
41   >>>
42   vector d: [ 0.  inf inf inf inf]
43   vector d: [ 0.   8. inf inf  1.]
44   vector d: [ 0.   3. 15.   7.  1.]
45   vector d: [0. 3. 8. 6. 1.]
46   Distances: [0. 3. 7. 6. 1.]
47   Parents: {0: 0, 1: 4, 4: 0, 2: 3, 3: 1}
```

7.5 Floyd-Warshall Algorithm

The Floyd-Warshall Algorithm aims to find all shortest paths between all nodes in a graph which can have positive or negative weight edges. The main idea of this algorithm is to improve the shortest path between two nodes in a graph until an optimal path is discovered using dynamic programming This algorithm assumes that all cycles in the graph have positive total weights.

This algorithm is depicted in Algorithm 7.5 where the matrix $D[n, n]$ shows the current distance between two nodes u and v and the matrix $P[n, n]$ contains the first node on the current shortest path from a node u to a node v. Whenever a shortest path is searched between two nodes u and v, whether a path through any intermediate node w provides a shorter path than the current one is tested. If a shorter path is found through w, the distance of the new path is assigned in the matrix D and w is stored in matrix P. The initialization of the algorithm between lines 2–9 takes n^2 steps, and the main body which consists of three nested loops takes $O(n^3)$ steps resulting in a time complexity of $O(n^3)$ for this algorithm [1].

Algorithm 7.5 $FW\_Routing$

1: $S \leftarrow \emptyset$
2: **for all** $\{u, v\} \in V$ **do**   ▷ initialize
3:    **if** $u = v$ **then**
4:        $D[u, v] \leftarrow \emptyset, P[u, v] \leftarrow \perp$
5:    **else if** $\{u, v\} \in E$ **then**
6:        $D[u, v] \leftarrow w_{uv}, P[u, v] \leftarrow v$
7:    **else** $D[u, v] \leftarrow \infty, P[u, v] \leftarrow \perp$
8:    **end if**
9: **end for**
10:
11: **while** $S \neq V$ **do**
12:    **pick** w from $V \setminus S$
13:    **for all** $u \in V$ **do**   ▷ Execute a global w-pivot
14:        **for all** $v \in V$ **do**   ▷ Execute a local w-pivot at u
15:            $D[u, w] \leftarrow \min(D[u, v], D[u, w] + D[w, v])$
16:        **end for**
17:    **end for**
18:    $S \leftarrow S \bigcup \{w\}$
19: **end while**

Python Implementation

Floyd-Warshall algorithm can be implemented by directly transforming the pseudocode of Algorithm 7.5 to Python code as below. The output of this program is the shortest path matrix D shown for graph of Fig. 7.3. The two nested for loops results in $O(n^3)$ time complexity as in the original algorithm.

```python
###############################################################
#                 Floyd-Warshall Algorithm                    #
###############################################################

import numpy as np
import math
inf = math.inf

def Floyd_Warshall(T):

    n=len(T)
    p = {}                         # holds parents
    T_old = T.copy()

    for k in range(0,n):           # do n-1 steps
        for i in range (0,n):
            for j in range(0,n):
                if T_old[i,j] > T_old[i,k]+T_old[k,j]:
                    T[i,j] = T_old[i,k]+T_old[k,j]
                    p[j,i] = k
                # temp = T_old[i,k]+T_old[k,j]
                # T[i,j] = min(T_old[i,j],temp)
        T_old = T
    return T, p


if __name__ == '__main__':

    B = np.array([[0,8,inf,inf,1],
                  [8,0,7,3,2],
                  [inf,7,0,1,inf],
                  [inf,3,1,0,5],
                  [1,2,inf,5,0]])

    D, parents = Floyd_Warshall(B)
    print ("Distances:")
    print(D)
    print ("Parents:")
    print(parents)
>>>
Distances:
[[0. 3. 7. 6. 1.]
 [3. 0. 4. 3. 2.]
 [7. 4. 0. 1. 6.]
 [6. 3. 1. 0. 5.]
 [1. 2. 6. 5. 0.]]
```

```
47   Parents:
48   {(2, 0): 4, (3, 0): 4, (0, 2): 4, (4, 2): 3, (0, 3): 4,
49   (2, 4): 3, (2, 1): 3, (1, 2): 3, (1, 0): 4, (0, 1): 4}
```

7.6 Transitive Closure

The *transitive closure* of a graph $G = (V, E)$ is defined as the graph $G' = (V, E')$ with edge $(u, v) \in E'$ if there is a path between the vertices u and v in G. Thus, for a connected graph which has paths between every vertex pairs it has, its transitive closure is a complete graph. The *connectivity matrix* of a graph G is a matrix C with entries $C[i, j]$ having a unity value if there exists a path between vertices i and j in the graph G. Finding the connectivity matrix of a graph G is basically finding the adjacency matrix of its transitive closure. We will see other ways of finding the connectivity matrices of directed and undirected graphs in Chap.8.

Warshall's algorithm to find the transitive closure of a graph works similar to finding distances using Floyd-Warshall algorithm, however, logical *and* and logical *or* operations are used instead of multiplication and addition performed during normal matrix multiplication.

Python Implementation

Python implementation of this algorithm has three nested *for* loops and the adjacency matrix T passed to this procedure simply multiplies this matrix with itself using logical operators. A test is carried in line 8 to check whether all elements of T is 1 which means graph G is connected, otherwise G is not connected as returned to the caller along with the connectivity matrix. We test two graphs represented by adjacency matrices A and B in the main procedure. The first graph shown in Fig. 7.4 is not connected and the second one represented by matrix B is connected by forming the connection between vertices 3 and 6 as shown by a dashed line in the figure. The output is as expected and the running time of this algorithm is $O(n^3)$ due to three nested *for* loops.

Fig. 7.4 Sample graph to test Warshall's algorithm

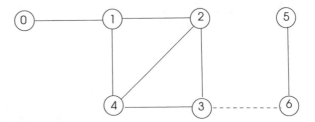

```
################################################################
#                 Warshal's Algorithm                         #
################################################################
1.    import numpy as np
2.    def Warshall(T):
3.        n=len(T)
4.        for k in range(0,n):              # do n^3 steps
5.            for i in range (0,n):
6.                for j in range(0,n):
7.                    T[i,j] = T[i,j] or (T[i,k] and T[k,j])
8.        if np.all((T == 1)):              # test T
9.            connected = True
10.       else:
11.           connected = False
12.       return T, connected

if __name__ == '__main__':
   A =np.array([[0,1,0,0,0,0,0],[1,0,1,0,1,0,0],[0,1,0,1,1,0,0],
        [0,0,1,0,1,0,0], [0,1,1,1,0,0,0], [0,0,0,0,0,0,1],
        [0,0,0,0,0,1,0]])
   D = Warshall(A)
   print ("Transitive Closure 1:")
   print(D.astype(int))
   B =np.array([[0,1,0,0,0,0,0],[1,0,1,0,1,0,0],[0,1,0,1,1,0,0],
        [0,0,1,0,1,0,1], [0,1,1,1,0,0,0], [0,0,0,0,0,0,1],
        [0,0,0,1,0,1,0]])
   D = Warshall(B)
   print ("Transitive Closure 2:")
   print(D.astype(int))
>>>
Transitive Closure 1:
[[1 1 1 1 1 0 0]
 [1 1 1 1 1 0 0]
 [1 1 1 1 1 0 0]
 [1 1 1 1 1 0 0]
 [1 1 1 1 1 0 0]
 [0 0 0 0 0 1 1]
 [0 0 0 0 0 1 1]]
G is not connected
Transitive Closure 2:
[[1 1 1 1 1 1 1]
 [1 1 1 1 1 1 1]
 [1 1 1 1 1 1 1]
 [1 1 1 1 1 1 1]
 [1 1 1 1 1 1 1]
 [1 1 1 1 1 1 1]
 [1 1 1 1 1 1 1]]
G is connected
```

7.7 Chapter Notes

We reviewed basic shortest path algorithms starting with the BFS algorithm in this chapter. The algebraic version of this algorithm is radically different than the classical one as it involves matrix-vector multiplication steps. The vector that contains the source is multiplied successively with the powers of the transpose of the adjacency matrix. The algebraic version has a much higher time complexity than the classical BFS algorithm, however, the matrix-vector multiplication can be easily parallelized using any of the methods described in Chap. 5. We then reviewed Dijkstra's and Bellman-Ford algorithms to find shortest paths in weighted graphs. These algorithms can be conveniently coded as algebraic algorithms as they use the distance matrix of a weighted graph as the input. Dijkstra's algorithm coded in Python needs $O(n^2)$ time as in the straightforward implementation without using special data structures of this algorithm. Bellman-Ford algorithm we implemented in Python requires $O(n^3)$ time.

The connectivity of a graph can be determined by Warshall's algorithm by finding the kth power of the adjacency matrix as Boolean multiplication where multiplications are performed by *and* operations and additions are done using *or* operations. The total time taken is $O(n^3)$ in Python implementation. Similarly, all-pairs-shortest-paths are found using Floyd-Warshall algorithm in $O(n^3)$ time. Matrix-matrix and matrix-vector multiplications are the key operations in all of these algorithms and these can be executed in parallel using any of the methods described in Chap. 5.

Exercises with Programming

1. Work out the adjacency matrix of the graph of Fig. 7.5 and then find the visited nodes at each level by implementing the Python BFS algorithm.
2. Find the shortest paths from the vertex 0 to all other vertices in the graph of Fig. 7.6 by running the Python Dijkstra's algorithm.
3. Provide all-pairs-shortest-paths in the graph of Fig. 7.7 by running the Python Floyd-Warshall algorithm.
4. Find the transitive closure of the graph of Fig. 7.8 by running the Python Warshall algorithm.

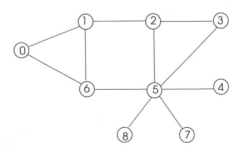

Fig. 7.5 Sample graph for Exercise 1

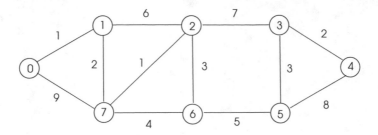

Fig. 7.6 Sample graph for Exercise 2

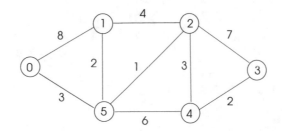

Fig. 7.7 Sample graph for Exercise 3

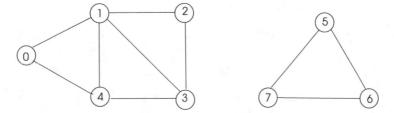

Fig. 7.8 Sample graph for Exercise 4

References

1. K. Erciyes, *Guide to Graph Algorithms: Sequential, Parallel and Distributed* (Springer, 2018)
2. J. Kepner, J. Gilbert, *Graph Algorithms in the Language of Linear Algebra* (2014)
3. K. Erciyes, *Distributed Graph Algorithms for Computer Networks*. Springer Computer Communications and Networks Series (2013)
4. D. Peleg, *Distributed Computing: A Locality-Sensitive Approach* (Monographs on Discrete Mathematics and Applications). Society for Industrial and Applied Mathematics (1987)

Connectivity and Matching

8

Abstract

A connected directed or undirected graph has paths between every pair of its vertices and an unconnected graph has more than one component. The connectivity of a computer network represented by a graph must be maintained for correct operation. We investigate algorithms to test connectivity and discovering the components of directed and undirected graphs in the first part of this chapter. We then look at the matching problem which is finding a disjoint set of edges in a graph. This problem is basically finding the maximum number of matching edges in an undirected graph and finding disjoint edges with a maximum/minimum possible total weight in a weighted graph.

8.1 Connectivity

Connectivity is an important property of a graph especially when the graph is used to represent a network. A connected graph provides paths between every pair of its vertices. Connectivity of a computer network needs to be maintained when nodes and links fail in such a network. We first review the basic theoretical concepts of connectivity and then describe algorithms to test connectivity, find connected components of undirected and directed graphs in this section.

8.1.1 Theory

Definition 8.1 (*connectivity*) An undirected or a directed graph is *connected* if there exists a path between any pair of its vertices. A graph G is said to be disconnected when there is at least one vertex pair (u, v) such that there does not exist a path between u and v in G.

© Springer Nature Switzerland AG 2021

K. Erciyes, *Algebraic Graph Algorithms*, Undergraduate Topics in Computer Science, https://doi.org/10.1007/978-3-030-87886-3_8

A graph G may not be connected, consisting a number of *components* which are connected subgraphs of G.

Definition 8.2 (*vertex-cut*) A *vertex-cut* of a connected graph $G = (V, E)$ is a subset V' of its vertices such that $G - V'$ has more components than G.

Informally, a graph G has more components when the vertices in V' with their adjacent edges are removed from G. If V' consists of a single vertex v, then v is called the *cut-vertex* of G. Note that a cut-vertex in a computer network is a single point of failure, thus, failing of such a node disconnects the network.

Definition 8.3 (*edge-cut*) An *edge-cut* of a connected graph $G = (V, E)$ is a subset E' of its edges such that $G - E'$ has more components than G.

If E' is a single edge e, then e is called the *bridge* of the graph G. Failure of a communication link represented by a bridge in a graph modeling the network leaves the network disconnected.

Definition 8.4 (*vertex connectivity*) The vertex-connectivity of a connected graph G is the minimum number of vertices removal of which results in a disconnected or trivial graph. This parameter is represented by $\kappa(G)$ for graph G.

Considering the computer network example, we would need $\kappa(G)$ to be as large as possible to tolerate node failures. The maximum value of this parameter is $n - 1$ in the case of a complete graph K_n where we would need to remove every adjacent vertex of a vertex to leave graph disconnected.

Definition 8.5 (*edge connectivity*) The edge-connectivity of a connected graph G is the least number of edges removal of which results in a disconnected graph. This parameter is represented by $\lambda(G)$ for graph G.

A graph G is called *k-edge-connected* if $\lambda(G) \geq k$. In a complete graph K_n, $\lambda(K_n) = n - 1$ and for every graph G of order n,

$$0 \leq \lambda(G) \leq n - 1 \tag{8.1}$$

These parameters are shown in Fig. 8.1 where vertices k, j, d and e are cut-vertices; edges (k, j), (d, j), (d, e), (e, f), (e, g), (e, h) are bridges; vertices b, l is one of the vertex-cuts, and edges (b, c) and (l, k) is one of the edge-cuts.

A block of a graph G is a subgraph G' of G such that G' does not contain any cut-vertex and G' is maximal. For example, the subgraph formed by vertices a, b and l in Fig. 8.1 is a block in this graph.

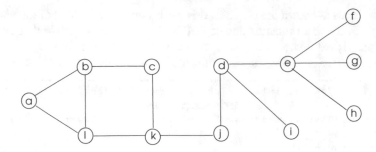

Fig. 8.1 A sample graph

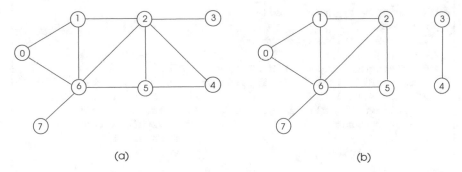

(a)   (b)

Fig. 8.2 A connected undirected graph

8.1.2 Undirected Graph Connectivity

Connectivity of an undirected graph and a digraph should be considered separately since connectivity of a digraph needs to be evaluated in both directions between a vertex pairs. We will review testing connectivity, finding the number and vertices of components of an undirected graph in this section.

8.1.2.1 Testing Connectivity in Python

The Breadth-First-Search (BFS) algorithm of Sect. 7.2 can be used to find whether an undirected graph is connected or not. The main idea here is that the BFS algorithm run from any node of the graph should visit all of its vertices. We can have an algorithm that calls the BFS algorithm which returns the identifiers of the vertices visited, and checks whether all of the vertices are visited by exactly one invocation of the BFS algorithm. We first test the connectivity of the undirected graph shown in Fig. 8.2a using this approach and find it is connected since BFS visits all vertices. This graph is modified to be disconnected as in (b) of the figure and when the adjacency matrix C of this graph is input to the algorithm, the output is displayed as not connected since vertices 3 and 4 are not visited by the BFS algorithm.

Python algorithm to test connectivity calls the function *Connectivity_Undirected* which calls the BFS algorithm that returns visited vertices in a list and sorts the visited vertices as shown below. Then, this list is compared with all of the nodes of the graph in the main function and connectivity is determined. Note that we need to sort the visited vertices *visited* to compare it with the sorted array of vertices *vertices*. However, sorting can be eliminated by using sets instead of lists.

```python
################################################################
#                 BFS-based Connectivity Test                 #
################################################################

import numpy as np
import BFS as bfs

def Connectivity_Undirected(A):
    n = len(A)
    vertices = np.array(range(n))
    s = 0
    N, visited, levels = bfs.BFS(A,s)    # call BFS
    print ("BFS visited vertices:", visited)
    visited.sort()                       # sort BFS visited
    if np.array_equal(visited,vertices): # compare visited
        return True                      # with all vertices
    else:                                # in order
        return False

if __name__ == '__main__':
    B = np.array([[1,1,0,0,0,0,1,0], [1,1,1,0,0,0,1,0],
        [0,1,1,1,1,1,1,0],[0,0,1,1,0,0,0,0],[0,0,1,0,1,1,0,0],
        [0,0,1,0,1,1,1,0],[1,1,1,0,0,1,1,1],[0,0,0,0,0,0,1,1]]
        , dtype=bool)
    C = np.array([[1,1,0,0,0,0,1,0], [1,1,1,0,0,0,1,0],
        [0,1,1,0,0,1,1,0], [0,0,0,1,1,0,0,0], [0,0,0,1,1,0,0,0],
        [0,0,1,0,0,1,1,0], [1,1,1,0,0,1,1,1], [0,0,0,0,0,0,1,1]],
        dtype=bool)
    if (Connectivity_Undirected(B)):
        print("B is connected")
    else:
        print("B is not connected")
    if (Connectivity_Undirected(C)):
        print("C is connected")
    else:
        print("C is not connected")
>>>
BFS visited vertices: [0, 1, 6, 2, 5, 7, 3, 4]
B is connected
BFS visited vertices: [0, 1, 6, 2, 5, 7]
C is not connected
```

Fig. 8.3 A sample graph
with three components

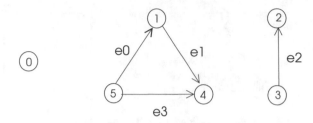

8.1.2.2 Finding Number of Components

We may be interested to find the number of components of a graph without actually
identifying these components. The algebraic properties of the graph provides this
information conveniently. The following theorems provide algebraic methods to find
the connectivity and components of a graph.

Theorem 8.1 *Let $B(G)$ be the incidence matrix of a connected graph with n vertices.
The rank of $B(G)$ is $n - 1$.*

Theorem 8.2 *Let $B(G)$ be the incidence matrix of a disconnected graph with k
components. The rank of $B(G)$ is $n - k$.*

Let us try to find the number of components of the graph of Fig. 8.3 using Python
with the stated property of the incidence matrix.

Sample Python code to find the number of connected components of a graph is
shown below. The imported Python library *numpy* has the function *linalg_matrix_rank*
which may be used to find the rank of a matrix. The graph of Fig. 8.3 has three components as the output from this program provides.

```
1   import numpy as np
2
3   def Comp_Num(A)
4       n = len(A)
5       rank=np.linalg.matrix_rank(A)
6       return n-rank
7
8   if __name__ == '__main__':
9       B = np.array([[0,0,0,0], [-1,1,0,0], [0,0,-1,0],
10          [0,0,1,0], [0,-1,0,-1], [1,0,0,1]])
11          n_comp = Comp_Num(B)
12          print ("Number of components of B:", n_comp)
13  >>> Number of components of B: 3
```

8.1.2.3 Finding Contents of Components

In order to find the identifiers of vertices in each component of a graph, we can
use the BFS algorithm again as follows. We run the basic algebraic BFS algorithm

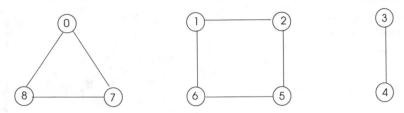

Fig. 8.4 A sample graph with three components

which returns the list of identifiers of vertices at each level. The contents of this list is copied to the new entry in the components and the visited vertices are removed from the graph. This process is repeated for each component until there are no unvisited vertices left. Let us implement this algorithm for the graph of Fig. 8.4 using Python.

The *Component_Find* algorithm in the following Python code calls the BFS algorithm which returns the visited vertices in visited which is copied to kth entry of dictionary *comps*. Each time we call the BFS algorithm, a new list is returned which is copied as a new entry to *comps*.

```
################################################################
#                 BFS-based Component Finding                  #
################################################################
import numpy as np
import BFS as bfs

def Component_Find(A):
    n = len(A)
    unmarked = [*range(0,n)]          # unmarked vertices
    comps = {}                        # holds components
    k = 0
    while len(unmarked) > 0:
        s = unmarked[0]
        N, visited, levels = bfs.BFS(A,s) # run BFS
        comps[k] = visited                # add new component
        for j in range(0,len(visited)):   # remove marked
            unmarked.remove(visited[j])
        k = k+1
    return comps

    if __name__ == '__main__':
        B = np.array([[1,0,0,0,0,0,0,1,1], [0,1,1,0,0,0,1,0,0],
            [0,1,1,0,0,1,0,0,0], [0,0,0,1,1,0,0,0,0],
            [0,0,0,1,1,0,0,0,0], [0,0,1,0,0,1,1,0,0],
            [0,1,0,0,0,1,1,0,0], [1,0,0,0,0,0,0,1,1],
            [1,0,0,0,0,0,0,1,1] ], dtype=bool)
        components = Component_Find(B)
        print ("Components:",components)
>>>
Components: {0: [0, 7, 8], 1: [1, 2, 6, 5], 2: [3, 4]}
```

8.1.3 Directed Graph Connectivity

A digraph has oriented edges and checking the existence of a path between two vertices must be done in both directions since having a path in one direction does not imply the other as in an undirected graph.

Definition 8.6 (*strongly connected digraph*) A digraph is said to be *strongly connected* if there exists a path in both directions for every u-v pair, from u to v and v to u.

```
1   import numpy as np
2
3   def Connectivity(A):
4       n = len(A)
5       C = A
6       for i in range(0,n):
7           C = np.dot(A,C)
8       C = C.astype(int)
9       return C
10
11  if __name__ == '__main__':
12      B =np.array([[1,0,0,0,0,0,0,1], [0,1,0,0,0,0,0,1],
13      [0,1,1,0,0,0,0,0], [0,0,1,1,0,1,0,0], [0,0,0,1,1,0,0,0],
14      [0,0,0,0,1,1,0,0], [0,0,1,0,0,0,1,0], [0,0,0,0,0,0,1,1]],
15      dtype=bool)
16      C = Connectivity(B)
17      print ("C = ",C)
18  >>>
19  C = [[1 1 1 0 0 0 1 1]
20   [0 1 1 0 0 0 1 1]
21   [0 1 1 0 0 0 1 1]
22   [0 1 1 1 1 1 1 1]
23   [0 1 1 1 1 1 1 1]
24   [0 1 1 1 1 1 1 1]
25   [0 1 1 0 0 0 1 1]
26   [0 1 1 0 0 0 1 1]]
```

8.1.3.1 Testing Strong Connectivity

Testing whether a digraph is strongly connected or not can be performed by a modified BFS algorithm. The procedure to follow is to record the visited vertices first, then run BFS on the transpose of the adjacency matrix and save the visited vertices for this case. Then, a comparison is made to check if these two sets are equal, if they do, then the digraph is strongly connected. We will illustrate this algorithm using Python for the graph of Fig. 8.5.

The algorithm *Strong_Connectivity* inputs the modified adjacency matrix A of the graph G and first calls the BFS algorithm to save the returned visited vertices in $vis1$

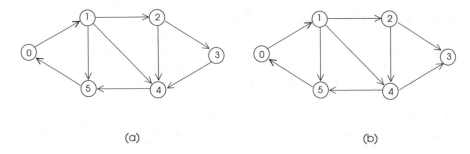

Fig. 8.5 Two sample digraphs

list. It then calls the BFS algorithm on the transpose of the adjacency matrix A and saves the visited list at $vis2$. Testing the equality of $vis1$ and $vis2$ shows whether G is strongly connected or not. Running this algorithm in the digraphs of Fig. 8.5 shows (a) is strongly connected and (b) is not strongly connected.

```
1   ##############################################################
2   #              BFS-based Strong Connectivity Test            #
3   ##############################################################
4   import numpy as np
5   import BFS as bfs
6
7   def Strong_Connectivity(A):
8       B = A.transpose()
9       N1, vis1, lev1 = bfs.BFS(A,0)
10      print ("BFS visited in G:",vis1)
11      N2, vis2, lev2 = bfs.BFS(B,0)
12      print ("BFS visited in G transpose:",vis2)
13      vis1.sort()
14      vis2.sort()
15      if vis1 == vis2:
16          return True
17      else:
18          return False
19
20  if __name__ == '__main__':
21      B =np.array([[1,1,0,0,0,0], [0,1,1,0,1,1], [0,0,1,1,1,0],
22          [0,0,0,1,1,0], [0,0,0,0,1,1], [1,0,0,0,0,1]], dtype=bool)
23      C =np.array([[1,1,0,0,0,0], [0,1,1,0,1,1], [0,0,1,1,1,0],
24          [0,0,0,1,0,0], [0,0,0,0,1,1], [1,0,0,0,0,1]], dtype=bool)
25      if Strong_Connectivity(B):
26          print ("B is strongly connected")
27      else:
28          print ("B is not strongly connected")
29      if Strong_Connectivity(C):
30          print ("C is strongly connected")
31      else:
32          print ("C is not strongly connected")
```

```
33   >>>
34   BFS visited in G: [0, 1, 2, 4, 5, 3]
35   BFS visited in G transpose: [0, 5, 1, 4, 2, 3]
36   B is strongly connected
37   BFS visited in G: [0, 1, 2, 4, 5, 3]
38   BFS visited in G transpose: [0, 5, 1, 4, 2]
39   C is not strongly connected
```

8.1.3.2 Strongly Connected Components

A strongly connected component of a digraph G is a subgraph G' of G such that G' is strongly connected, that is, there is a path between each vertex pair in G' in both directions. An algorithm to find SCCs of a digraph may be sketched as follows.

1. Find connectivity matrix C using the adjacency matrix A of the graph G.
2. Convert C to boolean.
3. Find transpose C_T of C.
4. Boolean multiply C and C_T.
5. The resulting product will have all 1s in the SCCs.

We will show the implementation of this algorithm for the graph of Fig. 8.6 using Python. The *Strongly_CC* inputs the adjacency matrix A of the graph and performs the above steps in the graph shown in Fig. 8.6 to find the three SCCs and store them in the dictionary *comps*.

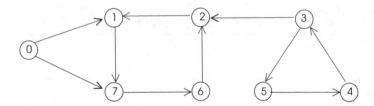

Fig. 8.6 A sample graph

```
1    ####################################################################
2    #                 Strongly Connected Components                   #
3    ####################################################################
4    import numpy as np
5    import CONNECT as con
6
7    def Strongly_CC(A):
8        n = len(A)
9        comps = {}          # strongly connected components
10       C = con.Connectivity(A)
11       C = C.astype(bool)  # convert C to boolean
```

```
12    C_T = C.transpose() # C transpose
13    SCC = np.logical_and(C,C_T) # logical and C with C_T
14    SCC = SCC.astype(int)   # convert SCC matrix to binary
15    k = 0
16    for i in range(0,n):
17        ones = np.where(SCC[i]==1)[0].tolist() # components
18        if ones not in comps.values():     # insert in comps
19            comps[k] = ones
20            k=k+1
21    return SCC, comps
22
23 if __name__ == '__main__':
24     B =np.array([[1,1,0,0,0,0,0,1], [0,1,0,0,0,0,0,1],
25         [0,1,1,0,0,0,0,0], [0,0,1,1,0,1,0,0],
26         [0,0,0,1,1,0,0,0], [0,0,0,0,1,1,0,0],
27         [0,0,1,0,0,0,1,0],[0,0,0,0,0,0,1,1]], dtype=bool)
28     SCC, components = Strongly_CC(C)
29     print ("SCC:")
30     print(SCC)
31     print ("SCCs:", components)
32 >>>
33 SCC:
34 [[1 0 0 0 0 0 0 0]
35  [0 1 1 0 0 0 1 1]
36  [0 1 1 0 0 0 1 1]
37  [0 0 0 1 1 1 0 0]
38  [0 0 0 1 1 1 0 0]
39  [0 0 0 1 1 1 0 0]
40  [0 1 1 0 0 0 1 1]
41  [0 1 1 0 0 0 1 1]]
42 SCCs: {0: [0], 1: [1, 2, 6, 7], 2: [3, 4, 5]}
```

8.2 Matching

Matching in a graph is a subset of its edges such that no two edge in this set is adjacent.

Definition 8.7 (*matching*) Given a graph $G = (V, E)$, a matching of G is the set $M \subset E$ such that any pair of edges $e_1, e_2 \in M$ do not share any endpoints.

In other words, a matching of a graph is an independent set of edges. As in the maximal independent set problem, we search a matching that is maximal (MM) which means it can not be enlarged any further. An unweighted maximum matching (MaxM) of a graph is a maximal matching with the greatest size among all matchings. Finding MaxM of a graph is one of the rare problems in graph theory that can be accomplished in polynomial time and there are various algorithms for this purpose.

We will be looking at basic algebraic matching algorithms in this section starting with the greedy ones.

8.2.1 Unweighted Matching

An unweighted graph $G = (V, E)$ does not have weights associated with its edges as noted. A simple strategy to find the maximal matching in a greedy way is to randomly select an edge in the graph, include it in the matching and remove all of its adjacent edges from the graph as in the steps below.

1. **Input**: An unweighted graph $G = (V, E)$
2. **Output**: A maximal matching M of G
3. $M \leftarrow \emptyset$
4. $E' \leftarrow E$
5. **while** $E' \neq \emptyset$
6.   randomly select $e \in E'$
7. $M \leftarrow M \cup \{e\}$
8. $E' \leftarrow E' \setminus \{e \cup$ all adjacent edges to $e\}$

We will implement this algorithm in Python; the function to find MM called *Maximal_Match* inputs the incidence matrix of the graph G with columns displaying edges that are incident to vertices in rows. The list M contains the edges in maximal matching and the list *unmarked* includes edges that can be matched. This algorithm iteratively selects an edge e at random from *unmarked*, includes e in matching and removes all adjacent edges to e from *unmarked* list as in the above steps.

```
################################################################
#            Random Unweighted Matching Algorithm            #
################################################################

import numpy as np
import random

def Maximal_Match(A):
    n=len(A)
    m=len(A[0])
    M = []
    unmarked = [*range(0,m)]    # unmatched edges
    while len(unmarked) != 0:
        r=random.randint(0,len(unmarked)-1) # select e
        e=unmarked[r]
        M.append(e)                  # include e in matching
        ends = np.argwhere(A[:,e]==1)
        ends.tolist()
        for j in range (0,m): # remove adjacent edges
```

```
20        if A[ends_list[0],j] == 1 and j in unmarked:
21            unmarked.remove(j)
22        if A[ends_list[1],j] == 1 and j in unmarked:
23            unmarked.remove(j)
24    return M
25
26 if __name__ == '__main__':
27    B = np.array([[1,0,0,0,0,0,0,0,0,1,0,1],
28    [1,1,0,0,0,0,0,0,1,0,0,0], [0,1,1,1,0,1,1,0,0,0,0,0],
29    [0,0,1,0,0,0,0,0,0,0,1,0], [0,0,0,1,1,0,0,0,0,0,0,0],
30    [0,0,0,0,1,1,0,1,0,0,0,0], [0,0,0,0,0,0,0,1,1,1,1,0,0],
31    [0,0,0,0,0,0,0,0,0,0,0,1,0], [0,0,0,0,0,0,0,0,0,0,0,1]])
32    MM = Maximal_Match(B)
33    print ("Maximal Matching:", MM)
34 >>>
35 Maximal Matching: [4, 11, 1, 10]
36 Maximal Matching: [2, 9, 4]
37 Maximal Matching: [10, 5, 8, 11]
```

The graph of Fig. 8.7 is used to test the Python algorithm and we can see the three sample outputs which show the identifiers of edges in matching are all valid maximal matchings for this graph.

8.2.2 Weighted Matching

A weighted graph has weights associated with its edges. Commonly, goal of a matching in such a graph is to find the matching with a minimum or maximum total weight of edges.

Definition 8.8 (*weighted matching*) Given a weighted graph $G = (V, E, w)$ and $w : E \rightarrow \mathbb{R}$, a maximum weighted matching MaxWM of G is a maximal matching of G with a maximum total weight among all maximal matchings of G.

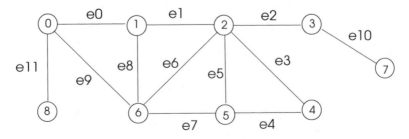

Fig. 8.7 An unweighted graph to test random matching algorithm

8.2.2.1 A Greedy Algorithm

A simple way to find MWM of a graph is to sort edges with respect to their weights in non-increasing order and to select edges from this list in order, include the edge in matching and remove all adjacent edges from the list as shown below.

1. **Input**: A weighted graph $G = (V, E, w)$
2. **Output**: A maximal weighted matching M of G
3. $M \leftarrow \emptyset$
4. sort E and store the list in Q
5. **while** $Q \neq \emptyset$
6.   remove first element e from Q
7. $M \leftarrow M \cup e$
8. $Q \leftarrow Q \setminus \{$ all adjacent edges to e $\}$

Let us try to implement this algorithm in Python with the above steps. Input to the Python function $Weighted\_Matching$ is the weighted incidence matrix of graph G. This matrix is similar to incidence matrix but we have weights of edges instead of 1s, similar to the relation between the adjacency matrix and the distance matrix. In order not to search all entries in this matrix for adjacent edges, we have the queue structure Q with entries as tuples $k, [i, j], w_k >$ where k is the edge identifier, i and j are the vertices this edge is incident and w_k is the weight of the edge. This structure is built and is sorted with respect to edge weights in lines 12–19. We then remove the first element from Q, include it in the matched edges set M and remove all of the adjacent edges to e from Q.

```
1   ###################################################################
2   #         Greedy Maximal Weighted Matching Algorithm             #
3   ###################################################################
4   import numpy as np
5
6   def Weighted_Matching(W):
7       n = len(W)
8       m = len(W[0])
9       M = []                       # matched edges
10      Q = []                       # queue
11      t_weight = 0
12      for j in range(0, m):        # build queue structure
13          ends = []
14          for i in range(0, n):
15              if W[i][j]!=0:
16                  ends.append(i)
17                  w = W[i][j]
18          Q.append([j,ends,w])
19      Q = sorted(Q, key=lambda x:x[2],reverse=True)
20      l = len(Q)
21      while l > 0:
22          e = Q[0]                 # get first Q element e
23          end1 = e[1][0]
```

```
24        end2 = e[1][1]
25        M += [e[0]]              # include e in matching
26        del Q[0]
27        t_weight = t_weight + e[2] # update total weight
28        k = 0
29        l1 = len(Q)
30        while k  < l1:          # delete adjacent edges
31            if Q[k][1][0] == end1 or Q[k][1][1] == end1:
32                del Q[k]
33                k = k-1
34            if Q[k][1][0] == end2 or Q[k][1][1] == end2:
35                del Q[k]
36                k = k-1
37            k = k+1
38            l1 = len(Q)
39        l = len(Q)
40    return M, t_weight
41
42 if __name__ == '__main__':
43    A = np.array(
44        [[12,0,0,0,0,0,0,0,0,0,0,7,0],[12,3,0,0,0,0,0,0,0,0,9,0,4],
45        [0,3,8,0,0,0,0,0,6,0,0,0,0],[0,0,8,10,0,2,0,3,0,0,0,0,0],
46        [0,0,0,10,5,0,0,0,0,0,0,0,0],[0,0,0,0,5,2,1,0,0,0,0,0,0],
47        [0,0,0,0,0,0,1,3,6,2,0,0,4],[0,0,0,0,0,0,0,0,0,2,9,7,0]])
48    m, total_weight = Weighted_Matching(A)
49    print("Maximal Weighted Matching Edges:",m)
50    print("Total Weight:", total_weight)
51 >>>
52 Maximal Weighted Matching Edges: [0, 3, 8]
53 Total Weight: 28
```

Running of this algorithm in the graph of Fig. 8.8 results in the correct outputs displayed.

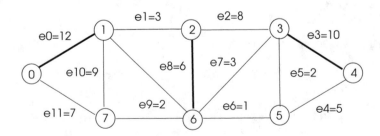

Fig. 8.8 A sample weighted graph

8.2.2.2 Algebraic Preis' Algorithm

Algorithm due to Preis provides a different way to solve the maximal weighted matching problem in a weighted graph. The algorithm consists of the following steps.

1. **Input**: A weighted graph $G = (V, E, w)$
2. **Output**: A maximal weighted matching M of G
3. $M \leftarrow \emptyset$
4. $E' \leftarrow E$
5. $V' \leftarrow V$
6. **while** $E' \neq \emptyset$
7.   select at random any $v \in V'$
8.   let $e \in E'$ be the heaviest edge incident to v
9. $M \leftarrow M \cup e$
10. $V' \leftarrow V' \{v\}$
11. $E' \leftarrow E' \setminus \{e$ and all adjacent edges to $e\}$

We will show two ways of implementing this algorithm in Python. In the first method, we will obey the steps of the algorithm outlined above. The distance matrix of the graph is input to the function $Preis\_Matching1$ which first forms the unmarked edge list um_edges as $[u, v]$ for each edge that has u and v as its endpoints. The unmarked vertices are kept in the list um_verts and a vertex v from this set is selected at random at each iteration of the algorithm. The maximum edge incident to this vertex is found and the other end of this edge is determined at lines 21–25. Since the um_edges list contains only one edge (u, v) with $u < v$, we need to order the found endpoints in lines 27–28. We now need to remove this edge and all other edges incident to endpoints and insert -1 for these edge entries so that they are not processed, again taking the ordering into account in lines 31–45. This process continues until there are no edges left in the list $unmarked\_edges$. Running of this algorithm in the weighted graph of Fig. 8.8 provides the sample outputs shown which are all valid maximal weighted matchings for this graph.

```
################################################################
            Preis' Maximal Weighted Matching Algorithm 1      #
################################################################
import numpy as np import random

def Preis_Matching1(A):
    n = len(A)
    m = len(A[0])
    M = []                        # matched edges
    um_edges = []                 # unmarked edges
    um_verts = list(range(0,n))   # unmarked vertices
    t_weight = 0                  # total weight
    for i in range(0,n):          # build list structure
        for j in range(i+1,n):
            if A[i,j] > 0:
```

```
16                  um_edges.append([i,j])
17       k = len(um_edges)
18       while k > 0:                  # do until all edges are processed
19           r = random.randint(0,len(um_verts)-1) # select a random
20           v = um_verts[r]                        # vertex v
21           m = max(A[v])             # find max edge on v
22           um_verts.remove(v)        # remove v from search
23           if m ==-1 or m==0:
24               continue
25           y = list(A[v]).index(m)
26           um_verts.remove(y)        # remove other end
27           a = min(v,y)              # order endpoints of v
28           b = max(v,y)
29           M.append([a,b])           # add edge to matching
30           t_weight = t_weight + m
31           for j in range(0,n):      # remove adjacent edges
32               if A[a,j] > 0:
33                   A[a,j] = -1
34                   A[j,a] = -1
35                   if a < j:
36                       um_edges.remove([a,j])
37                   else:
38                       um_edges.remove([j,a])
39               if A[b,j] > 0:
40                   A[b,j] = -1
41                   A[j,b] = -1
42                   if b < j:
43                       um_edges.remove([b,j])
44                   else:
45                       um_edges.remove([j,b])
46           k = len(um_edges)
47       return M, t_weight
48
49   if __name__ == '__main__':
50       D = np.array([[0,12,0,0,0,0,0,7],
51                     [12,0,3,0,0,0,4,9],
52                     [0,3,0,8,0,0,6,0],
53                     [0,0,8,0,10,2,1,0],
54                     [0,0,0,10,0,5,0,0],
55                     [0,0,0,2,5,0,1,0],
56                     [0,4,6,3,0,1,0,2],
57                     [7,9,0,0,0,0,2,0]])
58       m, total_weight = Preis_Matching1(D)
59       print("Maximal Weighted Matching Edges:",m)
60       print("Total Weight:", total_weight)
61   >>>
62   Maximal Weighted Matching Edges: [[2, 6], [0, 1], [3, 4]]
63   Total Weight: 28
64   Maximal Weighted Matching Edges: [[0, 1], [2, 3], [6, 7], [4, 5]]
65   Total Weight: 27
```

```
66    Maximal Weighted Matching Edges: [[1, 7], [4, 5], [2, 3]]
67    Total Weight: 22
```

In the second version of this algorithm, we explicitly label edges with integers as in the graph of Fig. 8.8. Now, the weighted incidence matrix is input to the function *Preis_Matching*2, and the list *unmarked* stores the edges that can be matched. Instead of selecting an unmarked vertex in the algorithm above, we select an edge e_i at random and check its endpoints. The heaviest edge e_j incident to the endpoints of e is then included in the matching, and all adjacent edges to e_j are removed from search. This process continues until there are no more edges left. Running of this algorithm in the sample graph of Fig. 8.8 resulted in the sample outputs shown.

```
1    ###############################################################
2    #          Preis' Maximal Weighted Matching Algorithm 2       #
3    ###############################################################
4
5    import numpy as np
6    import random
7
8    def Preis_Match(A):
9      n=len(A)
10     m=len(A[0])
11     M = []
12     unmarked = [*range(0,m)]    # unmarked edges
13     sum_w = 0
14
15     while len(unmarked) != 0:
16       r=random.randint(0,len(unmarked)-1) # select a random edge e
17       e=unmarked[r]
18       ends = np.argwhere(A[:,e]!=0)
19       max1=A[ends[0],0]
20       maxind1 = 0
21       for j in range(0,m):
22         if max1 < A[ends[0],j] and j in unmarked:
23             max1 =  A[ends[0],j]
24             maxind1 = j
25       max2=A[ends[1],0]
26       maxind2 = 0
27       for j in range(0,m):
28         if max2 < A[ends[1],j] and j in unmarked:
29             max2 =  A[ends[1],j]
30             maxind2 = j
31       if max1 > max2:
32           e  = maxind1
33           max_w = max1
34     else:
35           e = maxind2
36           max_w = max2
37       M.append(e)
38       sum_w = sum_w + max_w
```

```
39        ends = np.argwhere(A[:,e]!=0)
40        ends_list = [j[0] for j in ends.tolist()]
41        for j in range (0,m):
42            if A[ends[0],j] != 0 and j in unmarked:
43                    unmarked.remove(j)
44            if A[ends[1],j] != 0 and j in unmarked:
45                    unmarked.remove(j)
46     return M, sum_w
47
48  if __name__ == '__main__':
49
50     B = np.array([[12,0,0,0, 0,0,0,0,0,0,0,0,7],
51                   [12,3,0,0, 0,0,0,0,0,5,0,9,0],
52                   [0, 3,8,0, 0,0,0,0,6,0,0,0,0],
53                   [0, 0,8,10,0,2,3,0,0,0,0,0,0],
54                   [0, 0,0,10,5,0,0,0,0,0,0,0,0],
55                   [0, 0,0,0, 5,2,0,1,0,0,0,0,0],
56                   [0, 0,0,0, 0,0,3,1,6,5,2,0,0],
57                   [0, 0,0,0, 0,0,0,0,0,0,2,9,7]], dtype=int)
58
59     MM, sum_weight = Preis_Match(B)
60     print ("Total Weight of Matching:", sum_weight)
61     print ("Matched Edges:", MM)
62  >>>
63  Total Weight of Matching: [25]  Matched Edges: [3, 11, 8]
64  Total Weight of Matching: [28]  Matched Edges: [0, 8, 3]
65  Total Weight of Matching: [22]  Matched Edges: [11, 2, 4]
66  Total Weight of Matching: [28]  Matched Edges: [0, 3, 8]
67  Total Weight of Matching: [27]  Matched Edges: [2, 4, 0, 10]
```

8.2.2.3 Bipartite Graph Matching

A bipartite graph $G = (V_1 \cup V_2, E)$ has no edges between any two vertices in the vertex set V_1 and no edges between any two vertices in V_2. In other words, any edge in such a graph has one end in V_1 and the other end in V_2. Matching in a bipartite graph has the same goal as in a general graph; matched edges do not share any endpoints. As in a general graph, we can have unweighted and weighted matching in a bipartite graph. The greedy algorithms for unweighted and weighted matchings can be designed in a similar manner to the general graph case in a bipartite graph.

8.2.3 Rabin-Vazirani Algorithm

Tutte matrix T of an undirected graph $G = (V, E)$ is an n by n matrix with the following (i, j) entries,

$$T = \begin{cases} x_{i,j} & \text{if } (i, j) \in E, i < j \\ -x_{i,j} & \text{if } (i, j) \in E, i > j \\ 0 & \text{otherwise} \end{cases}$$

A perfect matching of a graph $G = (V, E)$ is $M \subseteq E$ such that every vertex of G is incident with exactly one edge of M. It can be shown that a graph G has a perfect matching if and only if $\det T \neq 0$ [?].

Rabin-Vazirani algorithm [?] is based on the following theorem.

Theorem 8.3 *Let $G = (V, E)$ be an undirected simple graph with a perfect matching and let T be its associated Tutte matrix. Then $T_{i,j}^{-1} \neq 0$ if and only if $G - \{i, j\}$ has a perfect matching.*

The following algorithm uses this theorem to find a perfect matching of a graph G by recursively removing edges from G.

1. **Input**: An undirected graph $G = (V, E)$
2. **Output**: A perfect matching M of G
3. $M \leftarrow \emptyset$
4. **while** $G \neq \emptyset$
5.   Compute T_G and form each variable with a random variable from $\{1, .., n\}$
6.   Compute T_G^{-1}
7.   Find i, j such that $(v_i, v_j) \in G$ and $T_G^{-1} \neq 0$
8. $M \leftarrow M \cup \{(v_i, v_j)\}$
9. $G \leftarrow G - \{v_i, v_j\}$
10. Return M

This algorithm finds a perfect matching with constant probability. It takes $O(n^{\omega+1})$ time since matrix inversion takes $O(n^\omega)$ time which is the dominant time.

8.3 Chapter Notes

We reviewed two fundamental concepts of graph theory in this chapter; connectivity and matching. There exists a path between any pair of vertices in a connected graph. Testing whether a graph is connected is needed in various applications, for example, to test whether a computer network is working. A cut-point vertex and a cut-point edge are vulnerable regions of a computer network and searching such areas may be needed to improve network resilience. A disconnected graph consists of two or more components. Connectivity in a digraph needs to be evaluated in both directions from a vertex pair. A digraph is said to be strongly connected if there is a path between each pair of vertices in both directions. A digraph may have strongly connected components with each component being strongly connected. We reviewed basic algorithms to find whether a graph is connected, finding its components, testing strong connectivity of a digraph and finding its strongly connected components commonly using the BFS algorithm of Chap. 7 and implemented all of these algorithms in Python.

A matching in a graph is a subset of its edges that are disjoint, that is, they do not share any endpoints. We reviewed a greedy algorithm to find a maximal matching in an unweighted graph and implemented this algorithm in Python in a sample graph. Weighted matching considers the weights of edges and a simple algorithm that always selects the largest weight edge can be designed as we described. Algorithm due to Preis takes a different approach by selecting a vertex and including the heaviest edge incident to that vertex in matching. We showed the implementation of both algorithms in Python and also provided a second version of Preis' algorithm.

The algorithms in Python have significant time complexities usually $O(n^2)$ or more, but there is always the chance of implementing these algorithms in parallel simply by implementing matrix multiplications in parallel.

Exercises with Programming

1. Show that a simple graph G of order n with $n \geq 2$ is complete if and only if its vertex connectivity number is $n - 1$.
2. Test whether the graph of Fig. 8.9 is connected or not using the Python BFS based algorithm and find the number of components of this graph using its rank. Provide the identifiers of vertices in each component using the BFS-based component finding Python algorithm.
3. Find whether the digraph of Fig. 8.10 is connected or not and output the vertices in each strongly connected component using the Python algorithms of Sect. 8.1.3.2.
4. Find the maximal weighted matching of the weighted graph of Fig. 8.11 using Python greedy weighted matching algorithm by first labeling the edges with identifiers.
5. Work out the maximal weighted matching of the weighted graph of Fig. 8.12 using Python Preis algorithms both versions by first labeling the edges with identifiers.

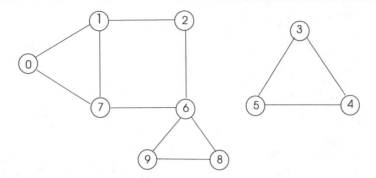

Fig. 8.9 A sample graph for Exercise 1

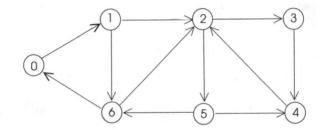

Fig. 8.10 A sample graph for Exercise 2

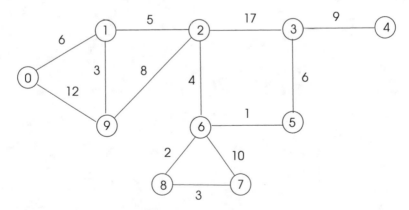

Fig. 8.11 A sample graph for Exercise 3

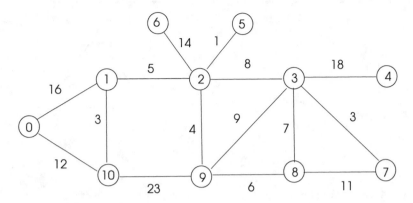

Fig. 8.12 A sample graph for Exercise 4

Subgraph Search

9

Abstract

A subgraph of a graph contains some of its vertices and edges. Some subgraphs have certain properties that can be used by many diverse applications. In this chapter, we survey a number of special subgraphs, algorithms on how to detect them using graph algebraic properties.

9.1 Independent Sets

An independent set of a graph consists of vertices that are not neighbors.

Definition 9.1 (*independent set*) An independent set of a graph $G = (V, E)$ is the set $V' \subset V$ such that for any $u, v \in V'$, edge $(u, v) \notin E$.

An independent set is maximal if it is not contained in any other independent set. The maximum independent set (MaxIS) of a graph G has the maximum order among all independent sets of G. The decision version of finding MaxIS is an NP-Complete problem [2] and thus, various heuristics are used to find maximal independent sets (MIS). The vertices of an MIS and a MaxIS of a sample graph are shown in Fig. 9.1a, b respectively as double circles.

9.1.1 A Greedy Algorithm

A simple greedy algorithm to find MIS I of a graph consist of the following steps.

1. Input: adjacency matrix A of $G = (V, E)$
2. Output: MIS I of G

© Springer Nature Switzerland AG 2021

K. Erciyes, *Algebraic Graph Algorithms*, Undergraduate Topics in Computer Science, https://doi.org/10.1007/978-3-030-87886-3_9

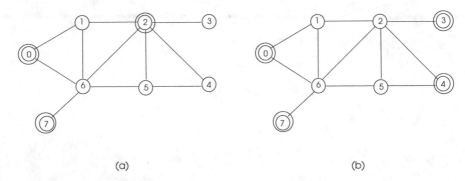

Fig. 9.1 MIS and MaxIS of a sample graph

3. $I \leftarrow \varnothing$
4. $V' \leftarrow V$
5. **while** $V' \neq \varnothing$
6.   randomly select a vertex $v \in V'$
7. $I \leftarrow I \cup \{v\}$
8. $V' \leftarrow V' \setminus \{v \cup N(v)\}$

Initially the set V' contains all vertices of G, we then select a random vertex v from this set, include it in the MIS set I and remove v and all of its neighbors from V'. This process continues until V' becomes empty. The resulting set I is an independent set since we obey independent set property by removing the neighbors of any selected vertex. It is maximal since V' not being empty means we can still include some more vertices in the independent set. This algorithm has a time complexity of $O(n^2)$ as the *while* loop runs $O(n)$ time for n vertices and removing adjacent vertices to the selected vertex also needs $O(n)$ time.

Python Implementation

Implementation of this algorithm in Python is shown below. Input to the function *MIS_Random* is the adjacency matrix of the graph of Fig. 9.1. Note the use of the variable k as we do not know beforehand how many neighbors of the selected vertex will be removed. The set I for three runs of the algorithm output is listed and we can see $\{3, 1, 5, 7\}$, $\{0, 3, 7, 4\}$ and $[2, 7, 0]$ are the two maximum independent sets for this graph.

```
1   ###############################################################
2   #        A Greedy Random Maximal Independent Set  Algorithm    #
3   ###############################################################
4   import numpy as np
5   import random
6
```

```
7   def MIS_random(A):
8     I = list()                # independent set
9     n = len(A)
10    verts = list(range(n))    # vertices 0,..., n-1
11    k = n
12
13    while k > 0:
14       r = random.randint(0,len(verts)-1) # select v
15       v = verts[r]
16       I.append(v)            # add it to MIS
17       k=k-1
18       verts.remove(v)
19       for j in range(0, n): # remove adjacent vertices
20          if A[v][j]!=0:
21             if j in verts:
22                verts.remove(j)
23                k = k-1
24    return I
25
26  if __name__ == '__main__':
27     B =np.array([[0,1,0,0,0,0,1,0], [1,0,1,0,0,0,1,0],
28                  [0,1,0,1,1,1,1,0], [0,0,1,0,0,0,0,0],
29                  [0,0,1,0,0,1,0,0], [0,0,1,0,1,0,1,0],
30                  [1,1,1,0,0,1,0,1],[0,0,0,0,0,0,1,0]])
31     mis = MIS_random(B)
32     print ("MIS =", mis)
33  >>>
34  MIS: [3, 1, 5, 7]
35  MIS: [0, 3, 7, 4]
36  MIS: [2, 7, 0]
```

9.1.2 Lowest Degree First Algorithm

A simple algorithm to find a MDS of a graph may use the heuristic which always selects the lowest degree vertex v for a maximal output, adds v to MIS, and removes all adjacent vertices v from graph. This process continues until graph is left with no vertices as shown in the steps below.

1. Input: adjacency matrix A of $G = (V, E)$
2. Output: MIS I of G
3. $I \leftarrow \varnothing$
4. $V' \leftarrow V$
5. calculate degrees of vertices from A
6. sort degrees from lowest to highest into a queue Q
7. **while** $V' \neq \varnothing$
8.   Remove first vertex v from Q
9. $I \leftarrow I \cup \{v\}$
10. $V' \leftarrow V' \setminus \{v \cup N(v)\}$

Python Implementation

A Python algorithm that inputs adjacency matrix A of an undirected graph is shown
below. Degrees of vertices are calculated by summing the rows at line 10, and a list
$merged$ with vertex identifier, vertex degree pairs is formed at the following line
and this list is sorted from lowest to highest at line 13. The rest of the algorithm
removes the first vertex v from merged, appends it to the MIS and all neighbors of v
are removed from the $merged$ list. This process continues until list becomes empty.

```
1   ###################################################################
2   #   Lowest Degree First Maximal Independent Set  Algorithm        #
3   ###################################################################
4   import numpy as np
5
6   def MIS_LDF(A):
7     I = list()   # MIS set
8     n = len(A)
9     vertices = list(range(n))
10    degrees = A.sum(axis=1) # compute degrees
11    degg = degrees.tolist()
12    merged = [(vertices[i],degg[i]) for i in range(0,len(vertices))]
13    merged.sort(key=lambda x:x[1]) # sort degrees with vertex id
14    k = n
15    while k > 0:
16        v,deg = merged.pop(0) # get the first vertex
17        I.append(v)          # add it to MIS
18        k = k-1
19        for j in range(0, n): # remove adjacent vertices of vertex v
20            if A[v][j]!=0:
21                if (j,degrees[j]) in merged:
22                    merged.remove((j,degrees[j]))
23                    k = k-1
24    return I
25
26  if __name__ == '__main__':
27      B =np.array([[0,1,0,0,0,0,1,0], [1,0,1,0,0,0,1,0],
28                   [0,1,0,1,1,1,1,0], [0,0,1,0,0,0,0,0],
29                   [0,0,1,0,0,1,0,0], [0,0,1,0,1,0,1,0],
30                   [1,1,1,0,0,1,0,1],[0,0,0,0,0,0,1,0]])
31      mis = MIS_LDF(B)
32      print ("MIS =", mis)
33  >>>
34  MIS = [3, 7, 0, 4]
```

Running of this algorithm for the graph of Fig. 9.1 favoring lower identities when
there is a tie results in the MIS $I = \{3, 7, 0, 4\}$ which is the MaxIS for this graph
shown in (b). Sorting needs $O(n \log n)$ steps and the while loop runs in $O(n)$ time.

We also need to search for neighbors of the dequeued vertex v in $O(n)$ time. Thus, the total time complexity of this algorithms is $O(n^2)$ due to two nested loops.

9.1.3 Luby's Algorithm

Luby provided an algorithm that finds the MIS of a given graph in parallel [3]. In each round of the algorithm, current active vertices are assigned random numbers and the vertex v that has the minimum random number assigned to it is appended to the MIS I. The vertex v and all of its neighbors are then deleted from the graph and this process continues until there are no more vertices left to be searched as shown in the below steps.

1. Input: $G = (V, E)$
2. Output: MIS I of G
3. $I \leftarrow \varnothing$
4. $V' \leftarrow V$
5. **while** $V' \neq \varnothing$
6.   assign a random number $r(v)$ to each vertex $v \in V'$
7.   **for all** $v \in V'$ in parallel
8.   **if** $r(v)$ is minimum amongst all neighbors
9. $I \leftarrow I \cup \{v\}$
10. $V' \leftarrow V' \setminus \{v \cup N(v)\}$

This algorithm terminates in $O(\log n)$ rounds with good probability [3]. The steps of the algorithm between lines 7–10 can be performed in parallel which will provide a speedup.

Python Implementation

We can code this algorithm in sequential form in Python as below. The list $values$ holds vertex v, its assigned random number and the list of neighbors of v as each element. The list I is used to store MIS nodes which is returned to the caller and the list $unmarked$ contains the vertices that are to be searched to be included in the MIS. In each round of the algorithm, vertices in $unmarked$ are assigned some random numbers between 0 and 1 in lines 20–21. Then the vertex v with the minimum value of the set of the first element of $unmarked$ and its neighbors is discovered in lines 22–30 and this vertex is added to the set I. The rest of the code deals with removing v and its neighbors from the list $unmarked$. Running this algorithm for three times for the graph of Fig. 9.1 resulted in the maximal independent set outputs.

```python
1   ####################################################################
2                   Luby's Maximal Independent Set  Algorithm        #
3   ####################################################################
4   import numpy as np from random import seed from random import random
5
6   def MIS_Luby(A):
7     I = []
8     n = len(A)
9     seed = 1
10    values = []
11    unmarked = list(range(n))
12    for i in range(0,n):
13      neighs = []
14      for j in range(0,n):
15        if A[i,j] == 1:
16          neighs.append(j)
17      values.append([i,0,neighs])
18    k = len(unmarked)
19    while k > 0:                          # generate random
20        for i in range(0,k):
21          values[unmarked[i]][1] = round(random(),2)
22        min_v = values[unmarked[0]][1]    # find min of first
23        min_i = unmarked[0]
24        l = len(values[unmarked[0]][2])
25        for j in range(0,l):
26          t = values[unmarked[0]][2][j]
27          if min_v  > values[t][1] and t in unmarked:
28                min_v = values[t][1]
29                min_i = t
30        I.append(min_i)                   # add min to I
31        pos = unmarked.index(min_i)
32        l = len(values[unmarked[pos]][2]) # remove neighbors of min
33        l2 = len(unmarked)
34        j = 0
35        t = unmarked[pos]
36        while j < l2:
37            l2 = len(unmarked)
38            for x in range(0,l):
39                if unmarked[j] == values[t][2][x] and
40                      values[t][2][x] in unmarked:
41                  del unmarked[j]
42                  l2 = l2-1
43            j = j+1
44        pos = unmarked.index(min_i)       # remove min
45        del unmarked[pos]
46        l2 = l2-1
47        k = len(unmarked)
48    return I
49
50  if __name__ == '__main__':
51      B =np.array([[0,1,0,0,0,0,1,0], [1,0,1,0,0,0,1,0],
52                   [0,1,0,1,1,1,1,0], [0,0,1,0,0,0,0,0],
53                   [0,0,1,0,0,1,0,0], [0,0,1,0,1,0,1,0],
54                   [1,1,1,0,0,1,0,1],[0,0,0,0,0,0,1,0]])
55      mis = MIS_Luby(B)
```

```
56        print ("MIS:", mis)
57   >>>
58   MIS: [1, 3, 4, 7]
59   MIS: [6, 3, 4]
60   MIS: [1, 3, 5, 7]
```

9.2 Dominating Sets

A dominating set of a graph G is a subset of its vertices such that any vertex of G is either in this set or adjacent to a vertex in this set.

Definition 9.2 (*dominating set*) A dominating set set of a graph $G = (V, E)$ is the set $V' \subset V$ such that for any $u \in V$, either $u \in V'$ or $u \in N(v)$ where $v \in V'$.

A minimum dominating set (MinDS) of a graph G is a dominating set of G with the minimum order among all dominating sets of G. Decision version of finding minimum set of a graph is an NP-Complete problem [2]. A minimal dominating set (MDS) of a graph can not be decreased any further, that is, removing a vertex from this set distorts dominating set property of the set. Fig. 9.2 displays a MDS of order 5 in (a) and a MinDS of order 3 in (b) with dominating set vertices shown in double circles.

Span Algorithm

In search of a minimal dominating set of a graph G, a common sense approach would be to include the higher degree vertices first. However, this method will also include high degree neighbors of higher vertices which do not need to be in the dominating set since a neighbor is already in the set. The span-based algorithm makes use of colors of nodes as follows: any node included in the set is black, any node that has

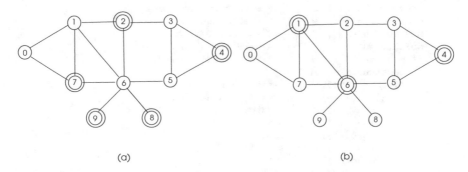

(a)   (b)

Fig. 9.2 MDS and MinDS of a sample graph

a black neighbor is colored grey and any other node is white. The span of a node is defined as the number of white neighbors it has including itself and the algorithm always selects the highest span node in the graph. Clearly, the algorithm should always run until all nodes are either black or grey to obey dominating set property.

Python Implementation

The Python code below implements this algorithm by first initializing each vertex color to white (line 8) and spans are calculated and stored at lines 10 and 11. Then, the *while* loop iterates until all nodes are colored black or grey by selecting the maximum span value vertex v at line 13, changing its color to black and modifying the neighbor colors and spans. This part of the code requires careful consideration based on the status of nodes as follows.

- Color of v is white and color of neighbor u is white: Node u is colored grey and its span is decreased twice due to itself and node v changing colors.
- Color of v is grey and color of neighbor u is white: The span of node u is decremented and its color is not changed.
- Color of v is grey and color of neighbor u is grey: Nothing is done.
- Color of v is grey and color of neighbor u is black: Nothing is done.

```
1    ######################################################################
2                  Span-based Minimal Dominating Set   Algorithm        #
3    ######################################################################
4
5    def MDS_Span(A):
6      MDS = []
7      n=len(A)
8      colors=['white']*n
9      unmarked = list(range(n))
10     degrees=A.sum(axis=1)
11     spans=degrees+1
12     while len(unmarked) > 0:      # check if any white nodes left
13         max_span = max(spans)      # find the maximum span node v
14         v = np.argmax(spans)
15         prev = colors[v]           # save previous color
16         colors[v] = 'black'        # change its color to black
17         MDS.append(v)              # include v in MDS
18         spans[v]=0                 # make its span 0
19         if prev == 'white':        # if unmarked, remove
20             unmarked.remove(v)
21         for j in range(0,n):       # check neighbors
22             if A[v][j]!=0:         # found a neighbor
23                 if prev == 'white' and colors[j] != 'black':
24                     spans[j] = spans[j]-1
25                 if colors[j] == 'white':
26                     spans[j] = spans[j]-1
27                     colors[j] = 'grey'
28                     unmarked.remove(j)
29                     for k in range(0,n):
```

```
30                        if A[j,k]==1 and colors[k] != 'black':
31                            spans[k] = spans[k]-1
32        return MDS
33
34    if __name__ == '__main__':
35        B =np.array([[0,1,0,0,0,0,0,1,0,0], [1,0,1,0,0,0,1,1,0,0],
36        [0,1,0,1,0,0,1,0,0,0], [0,0,1,0,1,1,0,0,0,0],
37        [0,0,0,1,0,1,0,0,0,0], [0,0,0,1,1,0,1,0,0,0],
38        [0,1,1,0,0,1,0,1,1,1], [1,1,0,0,0,0,1,0,0,0],
39        [0,0,0,0,0,0,1,0,0,0], [0,0,0,0,0,0,1,0,0,0]])
40        mds = MDS_Span(B)
41        print ("MDS =", mds)
42    >>>
43    MDS = [6, 3, 0]
```

Running of this algorithm in the graph of Fig. 9.2 results in the MDS $= \{6, 3, 0\}$. Note this set is another MinDS of this graph. The *while* loop runs in $O(n)$ time and checking the neighbors inside this loop is also $O(n)$ time resulting in $O(n^2)$ time in total.

9.3 Vertex Cover

A vertex cover of a graph is a set of its vertices such that any edge touches a vertex in this set. Finding vertex cover of a graph has various implementations such as building stores in an area such that any road reaches at least one store.

Definition 9.3 (*vertex cover*) A vertex cover of a graph $G = (V, E)$ (MinVC) is the set $V' \subseteq V$ such that any $(u, v) \in E$ has at least one endpoint in V'

A minimal vertex cover (MVC) is the set such that removing a vertex from this set leaves some edge or edges uncovered and the minimum vertex cover (MinVC) is the set with minimum order among all vertex covers of a graph. A MVC and an MinVC of a sample graph shown in double circles are depicted in Fig. 9.3. The decision version of finding MinVC of a graph is NP-Complete [3].

9.3.1 A Greedy Algorithm

A simple greedy algorithm to find MVC of a graph can be formed as follows.

1. Input: $G = (V, E)$
2. Output: MVC C of G
3. $VC \leftarrow \varnothing$
4. $E' \leftarrow E$
5. $V' \leftarrow V$

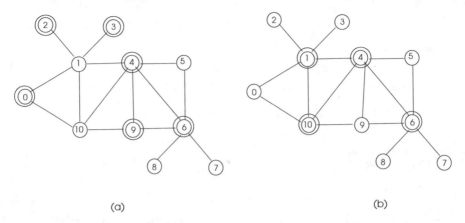

Fig. 9.3 MVC and MinVc of a sample graph

6. **while** $E' \neq \emptyset$
7.   randomly select a vertex $v \in V'$
8. $VC \leftarrow VC \cup \{v\}$
9. $E' \leftarrow E' \setminus$ any edge $(u, v) \in E$
10. $V' \leftarrow V' \setminus \{v\}$

Here, we start by equating the set E' to the edge set and V' to the vertex set of the graph. Then at each iteration, we select a vertex randomly from the set V', include it in the cover, remove all of its incident edges from E' and continue until E' becomes empty. Correctness is evident since any covered edge is removed from the graph and terminating condition is when all edges are covered. The output is a minimal vertex cover since we continue when there are uncovered edges meaning removing a vertex from the set VC leaves some edge or edges uncovered.

Python Implementation

This algorithm is implemented using Python as in the code below. The VC_Random procedure inputs the adjacency matrix of the graph and forms the degree array $degs$ which has the degrees of vertices at each entry and edge list $edges$ which contains all of the edges as (u, v) tuples in lines 12–18. A vertex v is then selected at random, appended to the vertex cover list VC and removed from the vertex list $verts$ in lines 20–23. All edges incident to v are then removed from the edge list in lines 25–29 Here, we also need to remove any vertex from the list that has all adjacent edges covered as otherwise we will have some redundant vertices in the cover, thus, the vertex cover will not be minimal. For this reason, the degree of each neighbor j of the randomly selected vertex v is decremented and if this degree becomes zero, vertex j is removed from the vertex list to be searched in lines 30–34. The whole process between lines 20 and 30 continues until the list *edges* becomes empty meaning all edges are covered by some vertex.

```
#################################################################
#           Random Greedy Minimal Vertex Cover Algorithm        #
#################################################################
import numpy as np
import random

def VC_Random(A):
  n=len(A)
  VC=[]
  degs = [0]*n
  verts = list(range(n))
  for i in range (0,n):
    degs[i] = np.sum(A[i],0)
  edges = []
  for i in range(0,n):      # form edges list
    for j in range(0,n):
        if A[i,j] == 1:
            edges.append((i,j))
  while len(edges) > 0:
    r = random.randint(0,len(verts)-1) # take a random vertex
    v = verts[r]becomes zero
    VC.append(v)
    verts.remove(v)
    k = len(edges)-1
    while k >= 0:                        # remove incident edges
        temp = edges[k]
        if temp[0] == v or temp[1] == v:
            edges.remove(temp)
        k = k-1
    for j in range(0,n):   # check if a neighbor is saturated
        if A[v,j] == 1:
            degs[j] = degs[j]-1
            if degs[j] == 0 and j in verts:
                verts.remove(j)
  return VC

if __name__ == '__main__':
    B = np.array([[0,1,0,0,0,1], [1,0,1,0,0,1], [0,1,0,1,1,1],
                  [0,0,1,0,1,0], [0,0,1,1,0,1], [1,1,1,0,1,0]])
    vc = VC_random(B)
    print ("Minimal Vertex Cover", vc)
>>>
Minimal Vertex Cover: [2, 4, 1, 0]
Minimal Vertex Cover: [5, 1, 3, 2]
Minimal Vertex Cover: [4, 5, 2, 1]
```

Running of this algorithm in the sample graph of Fig. 9.4 produced the sample outputs shown each of which is a MVC for this graph. The outer *while* loop runs

Fig. 9.4 A simple graph to run Python code

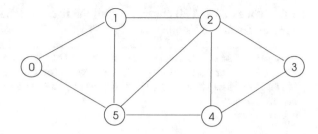

in $O(m)$ time for m edges, the inner *while* loop runs in $O(m)$ time and the inner *for* loop runs in $O(n)$ time resulting in a total time complexity of $O(nm)$ for this version of the algorithm.

9.3.2 A 2-Approximation Algorithm

An approximation algorithm for the vertex cover problem can be formed as follows. We find a maximal matching of the graph and include both ends of the matched edges in the vertex cover. We find matched edges because selecting and edge (u, v) for matching and including its endpoints in matching means covering edge (u, v) and all edges incident to u and v, thus, we do not need any adjacent edges. The steps of this algorithm may be formed as follows.

1. Input: $G = (V, E)$
2. Output: MVC C of G
3. $VC \leftarrow \varnothing$
4. $E' \leftarrow E$
5. $V' \leftarrow V$
6. **while** $E' \neq \varnothing$
7.   randomly select an edge $e = (u, v) \in E'$
8. $VC \leftarrow C \cup \{u, v\}$
9. $E' \leftarrow E' \backslash \{$any edge incident to u or $v\}$

Running of this algorithm in a simple undirected graph is depicted in Fig. 9.5. The randomly selected matching edges at each step are shown in bold. Both ends of selected edges are included in the vertex cover shown by dashed larger circles and the final vertex cover has all vertices of the graph except vertex 3 in (c). The minimum vertex cover for this graph has order 4 as depicted in (d) as vertices with solid double circles.

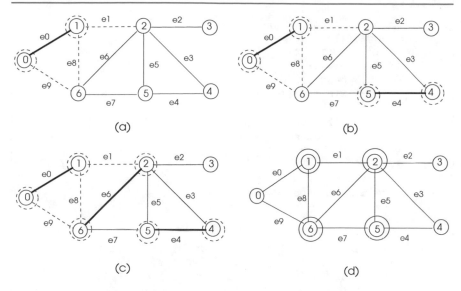

Fig. 9.5 Execution of 2-approximation algorithm in a simple undirected graph

Python Implementation

We will now describe the Python code for the 2-approximation algorithm. The main part of the code deals with deleting the randomly selected edge from the graph along with its neighboring edges. Input to the procedure *VC_Approx* is the incidence matrix $B(G)$ of the graph G. The unmarked edges are held in the list *edges* and the endpoints of a selected edge are included in the list VC at each iteration. The incidence matrix B of the graph of Fig. 9.5 is input and we find adjacent edges to the matched edge by iterating through rows of B. Each found edge is deleted from the list *edges* and main *while* loop iterates until this list becomes empty. We check adjacent edges to the matched edge from this matrix by deleting adjacent edges at each endpoint. The outputs of the algorithm for three runs are added to the end of the code.

```
###############################################################
#  A 2-Approximation Minimal Vertex Cover Algorithm           #
###############################################################

import random

def VC_Approx(B):
  n=len(B)
  m=len(B[0])
  VC=[]
  um_edges=list(range(m))
  k=len(um_edges)

  while k > 0:
```

```
15    r=random.randint(0,k-1)
16    e=um_edges[r]
17    for i in range(0, n):
18        if B[i][e]==1:
19            VC.append(i)
20            for j in range(0, m):
21                if B[i][j]==1:
22                    if j in um_edges:
23                        um_edges.remove(j)
24                        k=k-1
25        if k==0:
26            break
27    print ("Minimal Vertex Cover", VC)
28
29  if __name__ == '__main__':
30
31      B = ([[1,0,0,0,0,0,0,0,0,1], [1,1,0,0,0,0,0,0,1,0],
32          [0,1,1,1,0,1,1,0,0,0], [0,0,1,0,0,0,0,0,0,0],
33          [0,0,0,1,1,0,0,0,0,0], [0,0,0,0,1,1,0,1,0,0],
34          [0,0,0,0,0,0,1,1,1,1]])
35      VC_Approx(B)
36  >>>
37  Minimal Vertex Cover: [1, 6, 4, 5, 2]
38  Minimal Vertex Cover: [2, 4, 0, 6]
39  Minimal Vertex Cover: [0, 1, 4, 5, 2]
```

9.4 Coloring

Coloring in a graph refers to either coloring of its vertices, coloring of its edges or sometimes both. We will review the basic concepts, simple algorithms for these tasks and show the implementations of these algorithms using Python in this section.

9.4.1 Vertex Coloring

A vertex coloring of a graph is performed by assigning colors to its vertices such that each vertex has a color different than any of its neighbors.

Definition 9.4 (*vertex coloring*) Given a graph $G = (V, E)$, the vertex coloring of G is the function $\phi : V \to C$ which assigns colors to vertices in V from a set C of positive integers such that $\forall (u, v) \in E, \phi(u) \neq \phi(v)$.

In a graph with distinct vertex identifiers, say from 0 to $n - 1$, each identifier may correspond to a legal color. However, the *vertex coloring problem* is to color the vertices with minimum number of colors which is an NP-Complete problem in its decision version [2]. The chromatic number $\chi(G)$ of a graph G is the minimum vertex coloring number. The following pseudocode shows how to vertex color a graph based on some heuristic H.

1. **Input**: $G = (V, E)$, $C = \{0, 1, \ldots, k\}$
2. **Output**: Colors $C_0, C_1, \ldots, C_{n-1}$ of G
3. $V' \leftarrow V$
4. **while** $V' \neq \varnothing$
5.   Select a vertex $v \in V'$ based on heuristic H.
6.   Color v with the minimum color that does not conflict with the colors of its neighbors.
7. $V' \leftarrow V' \setminus v$

9.4.1.1 Highest-Degree First Algorithm

We can design an algorithm that uses the highest degree first heuristic as follows. We always select the highest degree uncolored vertex and color it with the minimum color that does not conflict with its neighbors.

Python Implementation

Coding in Python, we have the function to perform this task, $Vcol\_Hdeg$ which inputs the adjacency matrix A of the graph and initializes the array $vert\_colors$ to hold vertex colors. It then calculates the degree of each vertex and stores it in the array $degrees$.

The $while$ loop iterates until all vertices are colored and at each iteration, the vertex v with the highest degree is colored. The array $temp$ is used to find the current colors of the neighbors of vertex v and an unused smallest color is determined after sorting this array.

```
############################################################################
#           Highest Degree based Vertex Coloring Algorithm          #
############################################################################

import numpy as np

def Vcol_Hdeg(A):
    n=len(A)
    vert_colors = [n]*n              # holds colors of vertices
    colors = list(range(n))
    temp = [n]*n                     # temp array
    degrees=A.sum(axis=1)            # degrees of vertices
    count = n

    while count > 0:
        max_deg = max(degrees)       # find max degree vertex v
        v = np.argmax(degrees)
        degrees[v]=0
```

```
19        index=0
20        for j in range(0,n):          # find neighbors of v
21           if A[v][j]==1:
22              temp[index]=vert_colors[j] # copy neighbor colors
23              index = index+1
24        temp.sort()                   # sort temp
25        for k in range(0,n):          # find the smallest unused color
26           if temp[k] != k:
27              vert_colors[v]=k        # assign k to v
28              break
29        count=count-1
30     return vert_colors
31
32  if __name__ == '__main__':
33     B =np.array([[0,1,0,0,0,0,1,0,0,0,0],[1,0,1,0,0,1,1,0,0,0,0],
34                  [0,1,0,1,0,1,0,0,0,1,1],[0,0,1,0,0,1,0,0,0,0,0],
35                  [0,0,0,0,0,1,0,0,0,0,0],[0,1,1,1,1,0,1,0,0,0,0],
36                  [1,1,0,0,0,1,0,1,1,0,0],[0,0,0,0,0,0,1,0,0,0,0],
37                  [0,0,0,0,0,0,1,0,0,0,0],[0,0,1,0,0,0,0,0,0,0,0],
38                  [0,0,1,0,0,0,0,0,0,0,0]])
39     vcolors = Vcol_Hdeg(B)
40     print ("Vertex Colors:", vcolors)
41  >>>
42  Vertex Colors: [0, 2, 0, 1, 1, 2, 1, 0, 0, 1, 1]
```

Running of this algorithm for the graph of Fig. 9.6 provides the shown output which is a legal vertex coloring of this graph. Time complexity of this algorithm is $O(n^2)$ due to nested for loops inside the $while$ loop.

9.4.1.2 Vertex Coloring Using Independent Set

An independents set of a graph $G = (V, E)$ is the set $V' \subset V$ such that any pair of vertices in V' are not adjacent as was noted in Sect. 9.1. Therefore, finding an MIS of a graph and coloring all of these vertices with the same color does not violate vertex coloring property. We can then remove the MIS vertices from the graph and find the

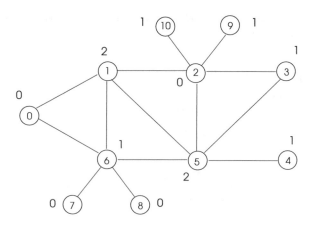

Fig. 9.6 A sample graph for highest-degree-first vertex coloring algorithm. Colors of vertices are shown next to them

MIS in the reduced graph and color these vertices with the next color and continue until all vertices are colored. Based on this observation, a vertex coloring algorithm based on MIS with the following steps is possible.

1. Input: $G = (V, E)$
2. Output: Minimal coloring $VCol$ of G
3. $color \leftarrow 0$
4. **while** $G \neq \varnothing$
5.   find maximal independent set I of G
6.   color all I vertices with $color$
7. $G \leftarrow G - I$
8. $color \leftarrow color + 1$

Python Implementation

We are now ready to code this algorithm in Python. The function $Vcol\_MM$ realizes the vertex coloring of the graph it receives in the form of adjacency matrix. We will be using the function MIS_LDF of module MIS to find maximal independent set of the graph using lowest degrees. The list $vert\_colors$ is used to store the colors of vertices and the list $verts$ holds the identifiers of uncolored vertices which is initialized to all vertices $0, \ldots, n - 1$. The $while$ loop runs until the length of uncolored vertex list becomes empty, thus, until all vertices are colored.

```
1    ####################################################################
2                 MIS based Vertex Coloring Algorithm           #
3    ####################################################################
4
5    import numpy as np
6    import MIS as mis
7
8    def Vcol_MIS(A):
9      n=len(A)
10     vert_colors = [-1]*n           # colors of vertices
11     verts = list(range(n))         # uncolored vertices
12     color = 0                      # initial color
13     while len(verts) > 0:          # do until all colored
14         I = mis.MIS_random(A)        # find MIS
15         for i in I:
16             vert_colors[verts[i]] = color # color vertices in MIS
17         A = np.delete(A,I,0)       # delete colored certices
18         A = np.delete(A,I,1)       # from adjacency matrix
19         I = sorted(I,reverse=True) # colored vertices
20         for i in I:
21             del verts[i]
22         color = color + 1          # next color
23     return vert_colors
24
25   if __name__ == '__main__':
26     B =np.array([[0,1,0,0,0,0,1,0,0,0,0],[1,0,1,0,0,1,1,0,0,0,0],
27                  [0,1,0,1,0,1,0,0,0,1,1],[0,0,1,0,0,1,0,0,0,0,0],
```

```
28                    [0,0,0,0,0,1,0,0,0,0,0],[0,1,1,1,1,0,1,0,0,0,0],
29                    [1,1,0,0,0,1,0,1,1,0,0],[0,0,0,0,0,0,0,1,0,0,0,0],
30                    [0,0,0,0,0,0,1,0,0,0,0],[0,0,1,0,0,0,0,0,0,0,0],
31                    [0,0,1,0,0,0,0,0,0,0,0]])
32         vcolors = Vcol_MIS(B)
33         print ("Vertex Colors:", vcolors)
34    >>>
35    Vertex Colors: [1, 0, 1, 0, 0, 3, 2, 0, 0, 0, 0]
36    Vertex Colors: [1, 0, 2, 0, 0, 1, 2, 0, 0, 0, 0]
37    Vertex Colors: [2, 0, 1, 0, 0, 2, 1, 0, 0, 0, 0]
```

Running of this algorithm three times in the graph of Fig. 9.6 provides the outputs shown which are legal colorings of vertices. Since the called MIS function randomly selects an independent set, the output may vary for different executions of the algorithm. The *while* loop runs $O(n)$ times for n vertices and the MIS_LDF algorithm has a $O(n^2)$ time complexity resulting in $O(n^3)$ time complexity in total for this algorithm.

9.4.2 Edge Coloring

Edge coloring refers to the coloring of the edges of a graph such that no two adjacent edges receive the same color.

Definition 9.5 (*edge coloring*) Edge coloring of a graph is the function $\phi'(G) = E \rightarrow C$ which assigns colors from the set $C = \{0, 1, \ldots, k\}$ to its edges such that $\phi'(e_i) \neq \phi'(e_2)$ if e_1 and e_2 are adjacent.

The *edge coloring problem* is to find the minimum number of colors to edge-color a given graph which is NP-Complete in its decision version [2]. The *edge chromatic number* $\chi'(G)$ of a graph G is the minimum number of colors to color G. We can see that this parameter is at least as large as the maximum degree of the graph since the vertex with the largest degree has to be colored with at least χ' colors, thus,

$$\chi'(G) \geq \Delta(G) \qquad (9.1)$$

Edge Coloring Using Matching

Matching in a graph was defined as the set of non-adjacent edges which means that edges in a matching can be colored with the same color. We can therefore design an algorithm based on this concept with the following steps

1. Input: $G = (V, E)$
2. Output: Minimal edge coloring of G
3. *color* $\leftarrow 0$
4. **while** $G \neq \varnothing$

5.   find maximal matching M of G
6.   color all M vertices with *color*
7. $G \leftarrow G - M$
8. $color \leftarrow color + 1$

Python Implementation

The Python program to implement this procedure is given below where the function
Ecol_MM finds the minimal edge colors for a graph input to it in the form of an
incidence matrix. Inside this function, we use the *Maximal_Match* function from
previously designed module MATCH to find the maximal matching in the graph. The
list *edge_colors* list holds the current edge colors and the list *edges* is used to hold
uncolored edges. The *while* loop runs until the *edges* list becomes empty meaning
all edges are colored. The graph is shrunk by removing the columns belonging to
the matched edges at each iteration.

```python
################################################################ #
Matching based Edge Coloring Algorithm                  #
################################################################

import numpy as np
import MATCH as match

def Ecol_MM(B):
   m=len(B[0])
   edge_colors = [-1]*m          # initial colors of edges
   edges = list(range(m))        # uncolored vertices
   color = 0                     # initial color
   while len(edges) > 0:         # do until all colored
      M = match.Maximal_Match(B) # find maximal matching M
      for i in M:
          edge_colors[edges[i]] = color # color edges in M
      B = np.delete(B,M,1)       # delete colored edge columns
      M = sorted(M,reverse=True) # remove colored edges
      for i in M:
            del edges[i]
      color = color + 1          # next color
   return edge_colors

if __name__ == '__main__':
I =np.array([[1,0,0,0,0,0,0,0,0,0,0,0,0,0,1],
[1,1,0,0,0,0,0,0,0,0,0,0,1,0],[0,1,1,1,1,1,0,0,0,0,0,0,0,0],
[0,0,1,0,0,0,1,1,1,0,0,0,0,0,0],[0,0,0,1,0,0,0,0,1,1,0,0,0,0,0],
[0,0,0,0,1,0,0,0,0,1,1,0,0,0,0],[0,0,0,0,0,1,0,0,0,0,1,1,1,1,1],
[0,0,0,0,0,0,0,0,0,0,0,1,0,0],[0,0,0,0,0,0,0,0,0,0,0,0,1,0,0],
[0,0,0,0,0,0,0,1,0,0,0,0,0,0],[0,0,0,0,0,0,1,0,0,0,0,0,0,0,0]])
   ecolors = Ecol_MM(I)
   print ("Edge Colors:", ecolors)
>>>
Edge Colors: [1, 2, 1, 5, 3, 4, 2, 0, 3, 0, 1, 2, 3, 0, 5]
```

```
35    Edge Colors: [0, 4, 2, 1, 0, 3, 1, 0, 3, 2, 4, 0, 1, 5, 2]
36    Edge Colors: [0, 2, 3, 0, 4, 1, 0, 1, 2, 1, 3, 0, 5, 4, 2]
```

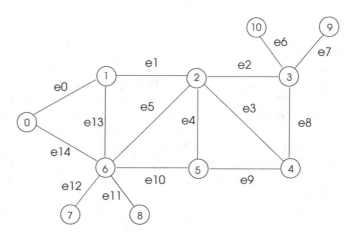

Fig. 9.7 A sample graph for edge coloring algorithm that uses matching

Running of this algorithm three times in the graph of Fig. 9.7 results in the outputs shown each of which is a legal coloring of the edges of this graph. Total number of colors is 6 from 0 to 5, and since $\Delta(G) = 6$ with vertex 6 in this graph, this result is logical in line with Eq. 9.1. Time complexity of this implementation is $O(mn^2)$ since *while* loop runs $O(m)$ times and the matching function has a time complexity of $O(n^2)$.

9.5 Chapter Notes

We reviewed some special subgraphs in this chapter as in [1]. An independent set of a graph is a subset of its vertices with no edge between any pairs of vertices in this set. The decision version of the problem of finding the independent set with the maximum order is NP-Complete. The second problem we investigated was the dominating set problem which searches for a subset of vertices of a graph such that any vertex of the graph is either in this set or neighbor to a vertex in this set. The k-dominating set problem extends this concept so as to have a dominating set where each vertex in the graph is either in this set or at most distance k from this set. Both versions of searching for a minimum order dominating set problem are NP-Hard.

We then reviewed the vertex cover concept and the algorithms to find the vertex cover of a graph which is again a subset of its vertices where each edge of the graph touches at least one vertex in this set. Finding such a set is NP-Hard but an approximation algorithm with a ratio of 2 can be designed using the matching of edges. The last problem we searched was coloring with two versions: vertex coloring and edge coloring. The vertex coloring of a graph is assigning colors in the form of integers to its vertices such that no two adjacent vertex receives the same color. The edge coloring similarly labels each vertex of a graph with colors such that each edge has a color label different than its neighbors. Finding the minimum number of colors in both cases is NP-hard.

Exercises with Programming

1. Improve Luby's MIS algorithm coded in Python by selecting a vertex at random, finding its neighbors and then finding the vertex with the minimum assigned random value to be included in the MIS.
2. Show the steps of the Spans algorithm to find minimal dominating set in the graph of Fig. 9.8
3. Implement the 2-Approximation vertex cover algorithm in the graph of Fig. 9.9 by showing each iteration.

Fig. 9.8 A sample graph for Exercise 1

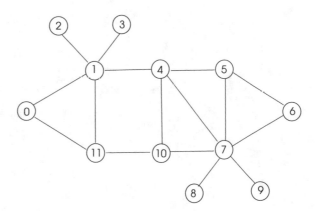

Fig. 9.9 A sample graph for Exercise 2

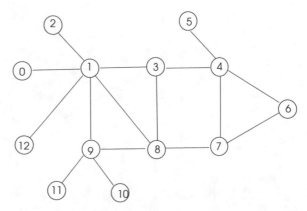

4. Design and implement an algorithm in Python that finds the minimal vertex coloring of a given graph using random selection.
5. Design and implement an algorithm in Python that finds the minimal edge coloring of a given graph using random selection.

References

1. K. Erciyes, *Distributed Graph Algorithms for Computer Networks*. Springer Computer Communications and Networks Series (2013)
2. M.R. Garey, D.S. Johnson, *Computers and Intractability* (W.H. Freeman, New York, 1979)
3. M. Luby, A simple parallel algorithm for the maximal independent set problem. SIAM J. Comput. **15**(4), 1036–1055 (1986)

Large Graph Analysis

10

Abstract

Large graphs consist of thousands of vertices and tens of thousands of edges between these vertices. Visualization and analysis of these graphs is difficult and new measures to evaluate structures of these graphs are needed. In this chapter, we review basic parameters to evaluate the structures of these graphs and conclude with main models of large networks.

10.1 Introduction

Large graphs consist of a huge number of vertices and edges. These graphs represent many real-life networks such as biological, technical and social networks. Analysis of such graphs is important since this provide information on their structure and behavior. For example, proteins inside the cell form a biological network called protein interaction network and analysis of such a network may provide useful information to determine the health and disease state of an organism.

The structure of the large graphs modeling real networks has some special properties; these graphs are not randomly connected. Instead, they commonly have very few vertices with many connections with the majority of the vertices having few connections, such networks are termed *scale-free networks*. Also, the distance between any two vertices in these real networks is very small compared to the number of nodes they have for which they are called *small-world networks*.

We start this chapter with basic parameters for the evaluation of these graphs and then continue with more advanced measures to identify structures of large graphs concluding with the main models of them.

© Springer Nature Switzerland AG 2021
K. Erciyes, *Algebraic Graph Algorithms*, Undergraduate Topics in Computer Science,
https://doi.org/10.1007/978-3-030-87886-3_10

10.2   Degree Distribution

The degree distribution of a graph shows the percentage of vertices with a given degree providing useful information about the structure of a graph.

Definition 10.1 (*degree distribution*) The degree distribution of a given degree k in a graph G is the ratio of the number of vertices with degree k to the total number of vertices.

The degree distribution displays the probability of a randomly selected vertex to have a degree k. Formally,

$$P(k) = \frac{n_k}{n} \tag{10.1}$$

Python program code to find the degree distribution of the graph in Fig. 10.1 and plot this distribution is given below. The *Degree_Dist* function inputs the adjacency matrix of a graph, finds the degrees of each vertex by multiplying this matrix with the unity vector U in lines 10–12. The number of vertices with the same degree is calculated and stored in vector C in lines 17–20. Each element of this vector is divided by the number of vertices to find $P(k)$ for $k = 0, ..., n - 1$ and finally this function is plotted against vertex degrees using the Python library *matplotlib.pylot*.

```
1    ###############################################################
2    #              Degree Distribution Algorithm                  #
3    ###############################################################
4
5    import numpy as np
6    import matplotlib.pyplot as plt
7
8    def Degree_Dist(A):
9      n=len(A)
10     U = np.full(n,1)
11     n=len(A)
12     D = np.dot(A,U)          # find degrees of vertices
13     C = [0]*n
14     k = 0
15     print(D)
16     D.sort()
17     for i in range(0,n): # calculate degree counts
18       if D[i] != k:
19          k = k+1
20       C[k] = C[k] + 1
21     P = [i/n for i in C]     # normalize
22     verts = list(range(n))
23     fig = plt.figure(figsize = (8, 5)) # plot
24     plt.bar(verts, P, width = 0.7)
25     plt.xlabel("Degrees")
26     plt.ylabel("Number of Vertices")
27     plt.title("Degree Distribution")
28     plt.xticks(np.arange(0, n-1, 1))
```

```
29        plt.yticks(np.arange(0,  0.45,  0.05))
30        plt.show()
31        return C
32
33   if __name__ == '__main__':
34
35        B = np.array([[0,1,0,0,0,0,0,0,1,0],
36                      [1,0,1,0,0,0,0,0,1,0],
37                      [0,1,0,1,0,0,0,1,0,0],
38                      [0,0,1,0,1,1,1,1,0,0],
39                      [0,0,0,1,0,0,0,0,0,0],
40                      [0,0,0,1,0,0,1,0,0,0],
41                      [0,0,0,1,0,1,0,1,0,0],
42                      [0,0,1,1,0,0,1,0,1,0],
43                      [1,1,0,0,0,0,0,1,0,0],
44                      [0,0,0,0,0,0,0,0,0,0]])
45
46        D = Degree_Dist(B)
47        print ("Degree Distribution:", D)
48   >>>
49   Degree Distribution: [1, 1, 2, 3, 1, 2, 0, 0, 0, 0]
```

Considering the graph of Fig. 10.1 again, the number of the vertices in this graph with degrees 0,...,9 is [1, 1, 2, 4, 1, 1, 0, 0, 0, 0]. Plotting of the percentage of vertices against the vertex degrees for this sample graph provides a visual display of degree distribution shown in Fig. 10.2. We can see that vertices with degree 3 are 40% of all with 2-degree vertices as 20% and the rest are 10% each.

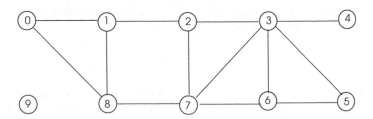

Fig. 10.1 An example graph for degree distribution

10.3 Density

The density of a graph is an important parameter that gives an overall idea about the structure of the graph. If we find that a graph is not dense, we may use data structures for sparse graphs for more efficient computations than done by dense graph structures such as adjacency or incidence matrices.

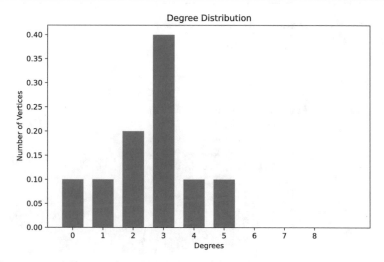

Fig. 10.2 Degree distribution of the graph of Fig. 10.1

Definition 10.2 (*graph density*) The density of a graph G denoted by $\rho(G)$ is the ratio of the number of its edges to the maximum possible number of edges that can exist in G as follows.

$$\rho(G) = \frac{2m}{n(n-1)} \tag{10.2}$$

The sum of degrees in an undirected graph G is $2m$ counting an edge from both ends and thus, the average degree of G, $deg(G)$, is $2m/n$. We can therefore rewrite using Eq. 10.2 as below.

$$\rho(G) = \frac{deg(G)}{(n-1)} \tag{10.3}$$

We can calculate the density of a graph using Eq. 10.2 in a Python program as below. We also find the density using Eq. 10.3 and print the outputs for the graph of Fig. 10.3.

```
1   ############################################################
2   #                    Density Algorithm                     #
3   ############################################################
4   import numpy as np
5
6   def Find_Density(A):
7       m = 0
8       n=len(A)
9       for i in range(0,n-1):      # finding density naive way
10          for j in range(i+1,n):
11              if A[i,j]= =1:
12                  m = m + 1
```

```
13    density1 = (2*m) / (n*(n-1))
14
15    S = np.zeros(n)              # finding density with numpy
16    for i in range(0,n):
17        S[i] = sum(A[i,:])
18    density2 = sum(S) / (n*(n-1))
19
20    return density1, density2
21
22  if __name__ == '__main__':
23
24    B = np.array([[0,1,0,0,0,0,1],
25                  [1,0,1,0,0,1,1],
26                  [0,1,0,1,1,1,0],
27                  [0,0,1,0,0,0,0],
28                  [0,0,1,0,0,0,0],
29                  [0,1,1,0,0,0,1],
30                  [1,1,0,0,0,1,0]])
31
32    Density1, Density2 = Find_Density(B)
33    print ("Densities", round(Density1,2), round(Density2,2))
34  >>>
35  Densities 0.43 0.43
```

Lastly, we find the density of the graph of Fig. 10.3 using the *numpy* built-in function *density* as shown below.

Fig. 10.3 A sample graph

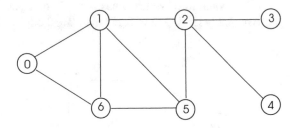

```
1   import networkx as nx
2   import pylab as plt
3
4   G = nx.Graph()
5   L = [(0,1),(1,2),(2,3),(0,6),(6,5),(1,5), (1,6),(2,5),(2,3),
6        (2,4)]
7   G.add_edges_from(L)
8   nx.draw(G, with_labels=True)
9   plt.savefig('graph.png')
10  print (" Density using graph = ",round(nx.density(G),2))
11  >>>
12  Density using graph =  0.43
```

10.4 Clustering Coefficient

The clustering coefficient of a vertex v in a graph is a parameter that reflects how well the neighbors of vertex v are connected.

Definition 10.3 (*clustering coefficient*) The clustering coefficient $CC(v)$ of a vertex v is the ratio of total number of edges between the neighbors of v to the maximum number of edges possible between these neighbors.

Let k be the number of neighbors of a vertex v in a graph G. Then, the maximum possible number of edges connecting vertices in $N(v)$ is $k(k-1)/2$. Therefore, the clustering coefficient of v can be expressed as follows.

$$CC(v) = \frac{2x}{k(k-1)} \tag{10.4}$$

where x is the existing number of connections of vertices in $N(v)$. The average clustering coefficient of a graph G, $CC(G)$, is the mean value of all of the clustering coefficients of vertices as stated below.

$$CC(G) = \frac{1}{n} \sum_{v \in V} cc(v) \tag{10.5}$$

Let us consider the graph of Fig. 10.4 to see how clustering coefficient of a vertex can be calculated. Vertex 0 has two neighbors, vertices 1 and 7, possible number of edges between them is one and since there are no edges connecting these vertices, $CC(0)$ is equal to 0. Vertex 3 has 4 neighbors, possible number of edges between these neighbors is $4 \cdot 3/2 = 6$ and there are 3 edges; (2,6), (4,5), (5,6), thus, $cc(3) =$

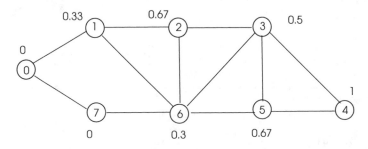

Fig. 10.4 A sample graph to calculate clustering coefficient of vertices

$3/6 = 0.5$. The rest of the clustering coefficients are calculated as shown next to them in the figure.

The Python program to calculate clustering coefficients of a graph and find the average clustering coefficient is shown below. The function *Clustering_Coeff* inputs the adjacency matrix of the graph G and a vertex v, and first finds the neighbors of v and stores them in the list N in lines 6–8. It then counts the number of edges between these vertices using the adjacency matrix and calculates the clustering coefficient of vertex v. The main function calls this procedure for all vertices and stores each result in dictionary CC which is used for output and also to find the average clustering coefficient of the graph. Running this algorithm for the graph of Fig. 10.4 provides the same outputs calculated by hand. This algorithm needs $O(n^2)$ operations due to nested for loops in lines 9–12.

```
1   def Clustering_Coeff(A,v):
2     N = []
3     n=len(A)
4     neighs = 0
5
6     for j in range(0,n): # find neighbors of v
7             if A[v,j]==1:
8                 N.append(j)
9     for i in range(0,len(N)): # count edges between them
10        for j in range(i+1,len(N)):
11            if A[N[i],N[j]]==1:
12                neighs = neighs + 1
13    n_neighs = sum(A[v,:])
14    if n_neighs == 0 or n_neighs == 1: # calculate CC(v)
15        cc = 0
16    else:
17        cc = (2*neighs)/(n_neighs * (n_neighs-1))
18    return cc
19
20  if __name__ == '__main__':
21
22      B = np.array([[0,1,0,0,0,0,0,1],
23                    [1,0,1,0,0,0,1,0],
24                    [0,1,0,1,0,0,1,0],
25                    [0,0,1,0,1,1,1,0],
26                    [0,0,0,1,0,1,0,0],
27                    [0,0,0,1,1,0,1,0],
28                    [0,1,1,1,0,1,0,1],
29                    [1,0,0,0,0,0,1,0]])
30      n = len(B)
31      CC = {}
32      for i in range (0,n):
33        CC[i] = Clustering_Coeff(B,i)
34        CC[i] = round(CC[i],2)
35      ave_cc = sum(CC) / n
36      print ("Clustering Coefficients")
37      print (CC)
```

```
38      print ("Average Clustering Coefficient:", round(ave_cc,2))
39   >>>
40   Clustering Coefficients
41   {0: 0.0, 1: 0.33, 2: 0.67, 3: 0.5, 4: 1.0, 5: 0.67, 6: 0.3,
42   7: 0.0}
43   Average Clustering Coefficient: 3.5
```

10.5 Matching Index

The matching index of two vertices in a graph is a parameter reflecting their similarity which is measured in terms of their common neighbors.

Definition 10.4 (*matching index*) The matching index of vertices u and v in a graph is the ratio of the number of their common neighbors to the union of all of their neighbors.

Evaluation of this parameter for vertices of graph of Fig. 10.5 yields the following: For vertices 0 and 1, their only common neighbor is 7, the union of their neighbors is {2, 7}, so their matching index is 0.5. The vertices 3 and 6 have two common neighbors as 2 and 5, the union of their neighbors is {2, 4, 5, 7}], so their matching index is also 0.5. All other pairwise matching indices can be calculated similarly. The matching indices for adjacent vertices is shown as labels of the edges joining these vertices in this graph.

The Python function to perform this task is called $Matching\_Index$ which inputs the adjacency matrix A and two vertices u and v of a graph. The lists N_u and N_v store neighbors of vertices u and v respectively in lines 13-22. Then, the list N which holds the union of the neighbors of u and v is formed and the number of common neighbors of these two vertices is calculated in lines 26-29 after which the matching index is formed and returned. Note that the vertices u and v are removed from the neighbor lists of each other in line with the definition of the matching index.

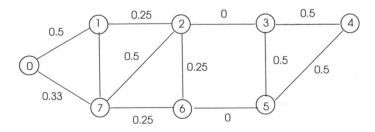

Fig. 10.5 A sample graph to find matching indices of vertex pairs

```
1   ################################################################
2   #                Matching Index Algorithm                      #
3   ################################################################
4
5   import numpy as np
6
7   def Matching_Index(A,u,v):
8     N_u = []
9     N_v = []
10    n=len(A)
11    n_common = 0
12
13    for j in range(0,n):      # find neighbors of u
14            if A[u,j]==1:
15                N_u.append(j)
16    if v in N_u:
17        N_u.remove(v)
18    for j in range(0,n):      # find neighbors of v
19            if A[v,j]==1:
20                N_v.append(j)
21    if u in N_v:
22        N_v.remove(u)
23    N = list(set().union(N_u,N_v)) # find their union
24    n_total = len(N)
25
26    for i in range(0,len(N_u)): # find common neighbors
27        for j in range(0,len(N_v)):
28            if N_u[i] == N_v[j]:
29                n_common = n_common + 1
30    match_index = n_common / n_total # calculate m.i.
31    return match_index
32
33  if __name__ == '__main__':
34
35      B = np.array([[0,1,0,0,0,0,0,1],
36                    [1,0,1,0,0,0,0,1],
37                    [0,1,0,1,0,0,1,1],
38                    [0,0,1,0,1,1,0,0],
39                    [0,0,0,1,0,1,0,0],
40                    [0,0,0,1,1,0,1,0],
41                    [0,0,1,0,0,1,0,1],
42                    [1,1,1,0,0,0,1,0]])
43      n = len(B)
44      M = np.zeros((n,n))
45      print ("Vertices and matching indices")
46      for i in range (0,n):
47          for j in range (i+1,n):
48              M[i,j] = Matching_Index(B,i,j)
49              M[i,j] = round(M[i,j],2)
50              print (i,"-",j, M[i,j]," ;   ", end='')
```

```
51  >>>
52  Vertices and matching indices
53  0 - 1 0.5 ;   0 - 2 0.5 ;   0 - 3 0.0 ;   0 - 4 0.0 ;
54  0 - 5 0.0 ;   0 - 6 0.25 ; 0 - 7 0.33 ; 1 - 2 0.25 ;
55  1 - 3 0.2 ;   1 - 4 0.0 ;   1 - 5 0.0 ;   1 - 6 0.5 ;
56  1 - 7 0.67 ; 2 - 3 0.0 ;   2 - 4 0.2 ;   2 - 5 0.4 ;
57  2 - 6 0.25 ; 2 - 7 0.5 ;   3 - 4 0.5 ;   3 - 5 0.33 ;
58  3 - 6 0.5 ;   3 - 7 0.17 ; 4 - 5 0.5 ;   4 - 6 0.25 ;
59  4 - 7 0.0 ;   5 - 6 0.0 ;   5 - 7 0.17 ; 6 - 7 0.25 ;
```

Running of this algorithm for the graph Fig. 10.5 provides the indices calculated by hand. The running time of this algorithm is $O(n^2)$ due to nested *for* loops. It is straightforward to parallelize this algorithm simply by sending the whole graph but have different vertex sets to be processed by each processor, for example, by sending vertices 0-99 to processor P_0, 100-199 to P_1, 200-299 to P_3 and 300-399 of a 400 node graph when indices are to be calculated by four processors P_0, P_1, P_2 and P_3.

10.6 Centrality

Centrality of a vertex or an edge of a graph are other parameters to evaluate the importance of a vertex or an edge. These parameters are commonly calculated by finding the number of neighbors of a vertex or the number of shortest paths through it and the number of shortest paths that pass through an edge.

10.6.1 Degree Centrality

The *degree centrality* of a vertex is simply its degree. This parameter shows the importance of a vertex, for example, a hub in a computer network will have a high degree. The average degree of a graph is simply the arithmetic average of the degrees of all vertices. It is difficult to have an idea on the overall structure of a graph based on this parameter only. The degree centralities of vertices can be found by multiplying the adjacency matrix with a vector of all ones as shown below,

$$DC = A \times U \qquad (10.6)$$

where $U = [1, 1, ..., 1]_n$ and $DC[i]$ is the degree centrality of vertex i. We will implement two Python functions to calculate degree centrality of the vertices of a graph. The first function $Degree\_Cent1$ adds the rows of the input adjacency matrix A to find the degrees and then sums the degrees. It lastly calculates the average degree of the input graph and returns the degree vector along with the average degree. In second function $Degree\_Cent2$, we make use of Eq. 10.6 and multiply matrix A with vector U of 1s, sum the product, calculate average degree and return the product and the sum. This function basically implements the same procedure as the fist function

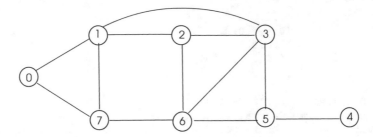

Fig. 10.6 A sample graph to calculate degree centralities of vertices

in matrix notation. The outputs from these two functions are then printed to give the same results for the sample graph of Fig. 10.6.

```python
##################################################################
#                 Degree Centrality Algorithm                   #
##################################################################

import numpy as np

def Degree_Cent1(A):
  n=len(A)
  D = [-1] * n
  sum_deg = 0
  for i in range(0,n):
    n_neighs = 0
    for j in range(0,n):
      if A[i,j] == 1:
        n_neighs = n_neighs + 1
    D[i] = n_neighs
    sum_deg = sum_deg + D[i]
  deg_ave = round(sum_deg/n,2)
  return D, deg_ave

def Degree_Cent2(A):
  n=len(A)
  U = np.full(n,1)
  n=len(A)
  D = np.dot(A,U)
  sum_deg = sum(D[:])
  deg_ave = round(sum_deg/n,2)
  return D, deg_ave

if __name__ == '__main__':

  sum_degree = 0
  B = np.array([[0,1,0,0,0,0,0,1],
                [1,0,1,1,0,0,0,1],
                [0,1,0,1,0,0,1,0],
```

```
                           [0,1,1,0,0,1,1,0],
                           [0,0,0,0,0,1,0,0],
                           [0,0,0,1,1,0,1,0],
                           [0,0,1,1,0,1,0,1],
                           [1,1,0,0,0,0,1,0]])

     C, deg_ave1 = Degree_Cent1(B)
     print ("Degree Centralities 1:", C)
     print ("Average Degree Centrality 1:", deg_ave1)
     E, deg_ave2 = Degree_Cent2(B)
     print ("Degree Centralities 2:", E)
     print ("Average Degree Centrality 1:", deg_ave2)
>>>
Degree Centralities 1: [2, 4, 3, 4, 1, 3, 4, 3]
Average Degree Centrality 1: 3.0
Degree Centralities 2: [2 4 3 4 1 3 4 3]
Average Degree Centrality 1: 3.0
```

10.6.2   Closeness Centrality

The closeness centrality $CC(v)$ of a vertex v in a graph is the reciprocal of the sum of the distances from v to all other vertices. This parameter basically shows how central a vertex is, a vertex that is close to all other vertices will have a large closeness centrality value than the others. Formally,

$$CC(v) = \frac{1}{\sum_{v \in V} d(u, v)} \tag{10.7}$$

with $d(u, v)$ denoting the distance between vertices u and v. In an unweighted graph, distances can be computed using the BFS algorithm, however, in a weighted graph, we need to find all-pairs-shortest-paths using Dijkstra's shortest path algorithm or Bellman-Ford algorithm of Sects. 7.3 and 7.4. In our Python implementation, we will assume an unweighted undirected graph and use BFS algorithm designed in Sect. 7.2. The function $Closeness\_Cent$ in the Python code below computes closeness centralities for all vertices of a graph given its adjacency matrix. The dictionary C is used to hold centrality values of vertices and the array D shows the distances among nodes in the end. For each vertex v of the graph, the BFS algorithm is run which returns the neighbor vertices of v in the matrix N which has i-hop neighbors in column i. Note that the BFS algorithm also returns the levels of nodes and vertices at each level which are not used in this algorithm. Then, we form the ith row of the distance matrix D by adding the transpose of the ith column of matrix N in line 15. For each vertex i, the sum of the ith row of matrix D is calculated to yield the closeness centrality value of i. This process is repeated for all vertices and the centrality vector C is returned with the matrix D.

```
################################################################
#              Closeness Centrality Algorithm                 #
################################################################

import numpy as np
import BFS as bfs

def Closeness_Cent(A):
    n = len(B)
    C = {}
    D = np.zeros((n,n), dtype=int)
    for i in range(0,n):
        N_i,v,l = bfs.BFS(A,i)
        for j in range(0,n):
            D[i,:] = D[i,:]+N_i.T[j,:]*(j)
        C[i] = sum(D[i,:])
        C[i] = round(1/C[i],2)
    return C, D

if __name__ == '__main__':

    B = np.array([[1,1,0,0,0,0,1,0],
                  [1,1,1,0,0,0,1,0],
                  [0,1,1,1,1,1,1,0],
                  [0,0,1,1,0,0,0,0],
                  [0,0,1,0,1,1,0,0],
                  [0,0,1,0,1,1,1,0],
                  [1,1,1,0,0,1,1,1],
                  [0,0,0,0,0,0,1,1]], dtype=bool)

    C,D = Closeness_Cent(B)
    print ("Distances:")
    print (D)
    print ("Closeness Centralities:")
    print (C)
>>>
Distances:
[[0 1 2 3 3 2 1 2]
 [1 0 1 2 2 2 1 2]
 [2 1 0 1 1 1 1 2]
 [3 2 1 0 2 2 2 3]
 [3 2 1 2 0 1 2 3]
 [2 2 1 2 1 0 1 2]
 [1 1 1 2 2 1 0 1]
 [2 2 2 3 3 2 1 0]]
Closeness Centralities:
{0: 0.07, 1: 0.09, 2: 0.11, 3: 0.07, 4: 0.07, 5: 0.09, 6: 0.11,
7: 0.07}
```

Fig. 10.7 A sample graph to
find closeness centrality

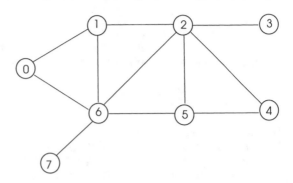

Running of this algorithm in the graph of Fig. 10.7 provides the shown distances
and centrality values. Considering vertex 0, it has distances of 1, 1, 2, 2, 3 and 3
to vertices 1, 6, 7, 2, 5, 3, 4 respectively for a total of 12 hops and its closeness
centrality is $1/12 = 0.07$. The running cost of this algorithm is not trivial, the BFS
algorithm runs in $O(n^2)$ time and it is called n times resulting in a total cost of
$O(n^3)$ operations. However, there are various opportunities in the code for parallel
processing. It is possible to run BFS independently for a number of disjoint vertex
sets by broadcasting the adjacency matrix. Calculation of the entries of vector C can
also be done in parallel.

10.6.3   Vertex Betweenness Centrality

A useful property of a vertex of a graph is the total number of shortest paths that pass
through it. This parameter called *vertex betweenness centrality* or just *betweenness
centrality* basically shows how important a vertex is, a vertex with a high value is on
more shortest paths than a vertex with a lower value. Formally,

$$BC(v) = \sum_{s \neq t \neq v} \frac{\sigma_{st}(v)}{\sigma_{st}} \tag{10.8}$$

where σ_{st} denotes the number of shortest paths between vertices s and t and $\sigma_{st}(v)$
is the number of paths that go through vertex v. Note that these values are calculated
for all s, t vertex pairs. The shortest paths for each vertex are calculated for a sample
undirected graph in Fig. 10.8. Note that we simply work out all modified BFS paths
considering multiple parents for nodes. For example, vertex 2 has two paths to reach
vertex 0 as (2-1-0) and (2-5-0) in (a) of this figure. Thus, the weight of edges in these
routes should be 0.5 each.

The shortest paths can be gathered in the following matrix where row i displays
the shortest path vertex weights to vertex i through every other vertex j in column
j. Note that when there are k shortest paths between two vertices u and v, we need
to weigh each vertex between u and v by $1/k$. For example, the weight of vertex 5 in
(a) of this figure is 1 for (3-4-5-0) route, 1 for (4-5-0) route, 0.5 for being in one of
the two routes (2-1-0) and (2-5-0); totalling 2.5. As another example, the weight of

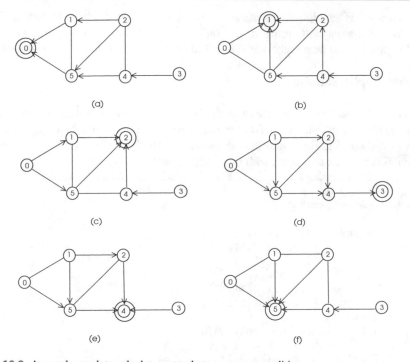

Fig. 10.8 A sample graph to calculate vertex betweenness centralities

vertex 4 in (d) is 1 for (0-5-4-3) route, 1 for (5-4-3) route, 1 for (2-4-3) route and 0.5 for being in one of the (1-2-4-3) and (1-5-4-3); for a total of 3.5. In (f), the weight of vertex 4 is 1.5 in total; 1 for (3-4-5) route and 0.5 for (2-4-5) route as there is an alternative (2-1-5) shortest path to vertex 5.

$$B = \begin{bmatrix} 0 & 0.5 & 0 & 0 & 1 & 2.5 \\ 0 & 0 & 1 & 0 & 1 & 1 \\ 0 & 0.5 & 0 & 0 & 1 & 0.5 \\ 0 & 0 & 0.5 & 0 & 3.5 & 1.5 \\ 0 & 0 & 0.5 & 0 & 0 & 1.5 \\ 0 & 0 & 0 & 0 & 1.5 & 0 \end{bmatrix}$$

Adding columns provides the total shortest paths through each vertex in the vector BC below,

$$BC = \begin{bmatrix} 0 & 1 & 2 & 0 & 8 & 7 \end{bmatrix}$$

Lastly, we need to divide these values by the total number of shortest paths which is $(n-1)(n-2)$ in an undirected graph counting paths in both directions, excluding the target node. This value is 20 for this graph with 6 nodes yielding vector VC as below,

$$VC = \begin{bmatrix} 0 & 0.05 & 0.1 & 0 & 0.4 & 0.35 \end{bmatrix}$$

Vertex 4 has the highest betweennnness value which can in fact be observed visually as it is the only cut-vertex of the graph being on the route of reaching vertex 3 and vertex 5 is the next highest as it is in the cross section of many shortest paths.

Python Implementation

Python *networkx* library provides a method called *betweenness centrality* which can be used to calculate this parameter for each vertex. The following code simply calls this method to find the centrality values for the graph of Fig. 10.6 and the values found this way are consistent with the values calculated by hand.

```
1    import numpy as np
2    import networkx as nx
3
4    A = np.array([ [0,1,0,0,0,1],
5                   [1,0,1,0,0,1],
6                   [0,1,0,0,1,1],
7                   [0,0,0,0,1,0],
8                   [0,0,1,1,0,1],
9                   [1,1,1,0,1,0]])
10   G = nx.Graph(A)
11   CV = nx.betweenness_centrality(G)
12   for key,value in CV.items():
13       CV[key] = round(CV[key],2)
14   print(CV)
15   >>>
16   {0: 0.0, 1: 0.05, 2: 0.1, 3: 0.0, 4: 0.4, 5: 0.35}
```

10.6.4 Edge Betweenness Centrality

The edge version of vertex betweenness centrality is the edge betwennes centrality, this time considering the ratio of the number of shortest paths through an edge to total number of shortest paths as shown below for edge e as formulated below.

$$BC(e) = \sum_{s \neq t \neq v} \frac{\sigma_{st}(e)}{\sigma_{st}} \tag{10.9}$$

Let us consider the graph of Fig. 10.6 again this time for edge betweenness values of edges in this graph. We follow a similar logic in calculation of these values to that of vertex betweenness values; if an edge (u, v) is on one of the k shortest paths from a vertex s to t, we label weight of (u, v) as $1/k$. For each shortest path, the edge weights are calculated and the final edge betweenness value of an edge is the sum of all its values for all shortest paths in the graph.

The shortest paths can be gathered in the following matrix EB where row i displays the shortest paths to vertex i through every other vertex j in column j.

$$EB = \begin{array}{c} \\ 0 \\ 1 \\ 2 \\ 3 \\ 4 \\ 5 \end{array} \begin{array}{cccccccc} (0,1) & (0,5) & (1,2) & (1,5) & (2,4) & (2,5) & (3,4) & (4,5) \\ \left(\begin{array}{cccccccc} 0.5 & 2.5 & 0.5 & 0 & 0 & 0.5 & 1 & 2 \\ 0 & 0 & 1.5 & 1 & 1 & 0 & 1 & 1 \\ 0.5 & 0.5 & 0.5 & 0 & 1 & 0.5 & 1 & 0 \\ 0 & 1.5 & 0.5 & 0.5 & 1.5 & 0 & 2.5 & 1.5 \\ 0 & 1 & 0.5 & 0.5 & 0.5 & 0 & 0 & 1.5 \\ 0 & 0 & 0 & 0 & 0 & 0 & 1 & 1 \end{array} \right) \end{array}$$

Adding columns provides the total shortest paths through each vertex in the vector EC below,

$$\begin{array}{cccccccc} (0,1) & (0,5) & (1,2) & (1,5) & (2,4) & (2,5) & (3,4) & (4,5) \\ (\quad 2 & 5.5 & 3.5 & 2.5 & 4 & 2 & 6.5 & 6.5 \quad) \end{array}$$

Lastly, normalization by dividing each element of this vector by $(n-1)(n-2)$ which is 20 for this graph yields vector BC,

$$\begin{array}{cccccccc} (0,1) & (0,5) & (1,2) & (1,5) & (2,4) & (2,5) & (3,4) & (4,5) \\ (\quad 0.1 & 0.23 & 0.17 & 0.13 & 0.2 & 0.1 & 0.33 & 0.33 \quad) \end{array}$$

Let us consider the graph of Fig. 10.9 again, this time for edge betwenness. We will calculate the edge betweenness values by the readily available Python method *edge_betweenness* from the *networkx* library as shown in the code below for the graph of Fig. 10.9. The returned value from this method is a dictionary with keys showing the edges and values showing the betweenness values which are consistent with what we calculated by hand.

```
import numpy as np
import networkx as nx

A = np.array([ [0,1,0,0,0,1],
               [1,0,1,0,0,1],
               [0,1,0,0,1,1],
               [0,0,0,0,1,0],
               [0,0,1,1,0,1],
               [1,1,1,0,1,0]])
G = nx.Graph(A)
CE = nx.edge_betweenness_centrality(G)
for key,value in CE.items():
    CE[key] = round(CE[key],2)
print(CE)
>>>
{(0, 1): 0.1, (0, 5): 0.23, (1, 2): 0.17, (1, 5): 0.13,
(2, 4): 0.2, (2, 5): 0.1, (3, 4): 0.33, (4, 5): 0.33}
```

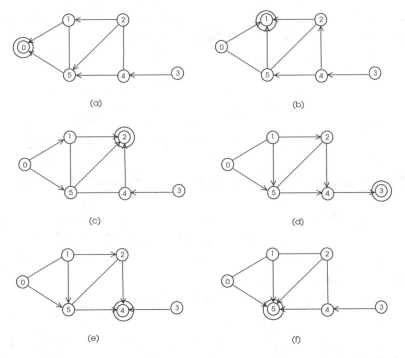

Fig. 10.9 A sample graph to calculate edge betweenness values

10.6.5   Eigenvalue Centrality

A node in a network may be connected to an important node although itself may not be important. This is indeed the case in a social network when reaching a well-known personality may be achieved by someone who is close to that person. The idea in eigenvalue centrality is to assign importance to a node based on the importance of its neighbors. This relation basically is extension of the degree centrality by considering degrees of the neighbors. This relation can be formalized by the following equation.

$$x_i = \frac{1}{\lambda} \sum_{j \in N(i)} x_j = \frac{1}{\lambda} \sum_{j \in v} a_{ij} x_j \qquad (10.10)$$

with $N(i)$ as neighbors of node i and λ is a constant. The lefthandside of this equation is the vector x and the righthandside is the product of the adjacency matrix A with this multiplied by the reciprocal of λ as $x = (1/\lambda)Ax$ which when organized as $Ax = \lambda x$ is the equation for matrix A. The adjacency matrix has n eigenvalues and n eigenvectors associated with thes values. There will be a unique eigenvalue by Perron-Frobenius theorem [2] and the corresponding eigenvector will contain the eigenvalue centralities of the nodes.

We can have the following procedure steps to find the eigenvalue centrality of the nodes of a graph.

1. **Input**: The adjacency matrix A of a graph
2. **Output**: The eigenvector X_m that contains eigenvalue centralities.
3. Let $\det(A - \lambda I) = 0$ and compute eigenvalues $\lambda_1, ..., \lambda_n$.
4. Find the largest eigenvalue λ_m of the eigenvalues.
5. Compute the eigenvector X_m related to λ_m

We will write the Python code for this procedure using the readily available procedure *eig* from the library *np.linalg* to find eigenvalues and eigenvectors of a given matrix in lines 10–11. The maximum eigenvalue and the eigenvector corresponding to this value is returned to the main program which prints the eigenvalue centralities stored in the vector.

```
##################################################################
                 Eigenvalue Centrality Algorithm                 #
##################################################################

import numpy as np

def Eigenvalue_Cent(A):

  n=len(A)
  eigvals, eigvecs = np.linalg.eig(A)
  eigvals = eigvals.real
  eigvecs = eigvecs.real
  print ("Eigenvalues:")
  for i in range (0,n):
     print (round(eigvals[i],2),end='  ')
  print ("\n\nEigenvectors:")
  for i in range (0,n):
    for j in range(0,n):
         eigvecs[i,j] = round(eigvecs[i,j],2)
  print(eigvecs)
  max_index = np.argmax(eigvals)
  return eigvecs[max_index]

if __name__ == '__main__':

  B = np.array([[0,1,0,0,0,0,0,1],
               [1,0,1,0,0,0,1,1],
               [0,1,0,0,1,1,0,0],
               [0,0,0,0,1,0,0,0],
               [0,0,1,0,0,1,0,0],
               [0,0,1,0,1,0,0,0],
               [0,1,0,0,0,0,0,1],
               [1,1,0,0,0,0,1,0]])
  E = Eigenvalue_Cent(B)
  print ("\nEigen Centralities: ", E)
>>>
Eigenvalues:
0.0   2.77   1.92   -1.89   -0.44   -1.36   -0.0   -1.0

Eigenvectors:
```

```
41   [[ 0.      0.37   0.21 -0.36   0.12   0.4   -0.64 -0.   ]
42    [ 0.      0.56   0.12   0.63 -0.38   0.16 -0.    -0.   ]
43    [ 0.      0.34 -0.47 -0.52 -0.4   -0.32 -0.     0.   ]
44    [ 1.      0.07 -0.27 -0.1  -0.64 -0.1  -0.43 -0.58]
45    [ 0.      0.19 -0.51   0.18   0.28   0.13   0.     0.58]
46    [ 0.      0.19 -0.51   0.18   0.28   0.13   0.    -0.58]
47    [ 0.      0.37   0.21 -0.36   0.12   0.4    0.64 -0.   ]
48    [ 0.      0.47   0.28   0.04   0.33 -0.71   0.     0.   ]]
49
50   Eigen Centralities:  [ 0.     0.56   0.12   0.63 -0.38   0.16 -0.
51     -0.   ]
```

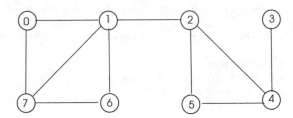

Fig. 10.10 A sample graph to test eigenvalue centrality

Running this algorithm in Fig. 10.10 results in the displayed the eigenvalue centralities. We can see that the node 3 has the highest centrality value as it is the only single node to access the network externally.

10.7   Network Models

The basic network models for large networks can be classified as random networks, small-world networks and scale-free networks.

10.7.1   Random Networks

The random graph model introduced by Erdos and Renyi [?] (ER), a graph is constructed by randomly adding edges to the graph with some probability p. The value of this parameter is independent for each edge. When $p = 1$, we have a complete graph. The Python *networkx* library has a function *erdos_renyi_graph* to draw an ER graph with probability p using the *matplotlib* library to visualize the plots as in the code below.

```
1   import networkx as nx
2   import matplotlib.pyplot as plt
3
4   P = [0.2, 0.5, 1]
5   for i in range(0,len(P)):
6       G=nx.erdos_renyi_graph(10,P[i])
7       nx.draw(G,with_labels=1)
8       plt.show()
```

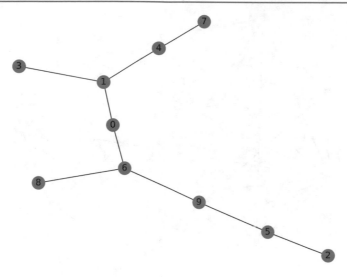

Fig. 10.11 Erdos-Renyi graph with $p = 0.25$

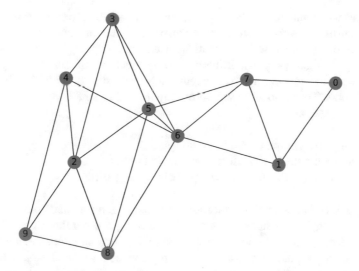

Fig. 10.12 Erdos-Renyi graph with $p = 0.5$

We draw random graphs with probabilities 0.25, 0.5 and 1 as shown in Figs. 10.11, 10.12 and 10.13.

10.7.2 Small-World Networks

Small world property of a network is manifested by a small maximum distance between any two nodes of a network. Historically, this property of social networks was observed in Milgram experiment [?] in which the average distance between two

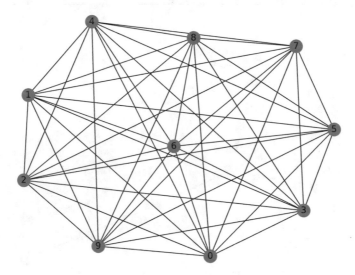

Fig. 10.13 Erdos-Renyi graph with $p = 1$

individuals was found to be 6 hops. Later, the small-world property was found to be valid in real-life networks such as protein interaction networks and Internet [1]. The mean distance in a small-world network approximates $O(\log(n))$ as $n \to \infty$.

Watts and Strogatz [4] analyzed this model and concluded the small world property is prevalent with a low average path length and high clustering coefficient in various large networks. They also proposed a procedure to build a small-world network with the following steps:

1. Construct a ring graph G that has n nodes.
2. Connect each node to its k closest neighbors to get G'.
3. Rewire each edge (u, v) in G' with probability $p \in (0, 1)$

A ring lattice is obtained by first forming a ring of n nodes and connecting each node to its neighbors 2 hops away, thereby forming a 4-regular lattice. For a 6-regular lattice, we need to connect each node to its 3-hop neighbors. A small value of p results in a regular lattice and a large p provides a graph close to ER graph. Selecting p between 0 and 1 would give the small-world network required. Python *networkx* library provides the function *watts_strogatz_graph* to build a small world network of given number of nodes, edges connected to each node and the probability of wiring the edges.

```
import networkx as nx
import matplotlib.pyplot as plt

G=nx.watts_strogatz_graph(10,5,0.5)
pos = nx.circular_layout(G)
nx.draw(G,pos,with_labels=True)
plt.show()
```

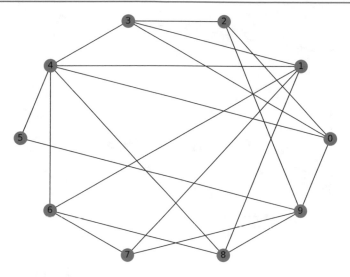

Fig. 10.14 Watts-Strogatz graph with $p = 0.5$

Using this function with 10 nodes and 5 edges with probability 0.5 resulted in the sample graph shown in Fig. 10.14 with a maximum distance of 3.

10.7.3   Scale-Free Networks

A scale free network has a degree distribution which follows power law. Formally, the fraction $P(k)$ of nodes of the nodes with degree k is expressed as,

$$P(k) = ck^{-\gamma} \tag{10.11}$$

with γ commonly having a value between 2 and 3 and c is a constant. This model proposed by Barabasi and Albert relies on two main concepts; *growth* and *preferential attachment*. Growth concept assumes that the number of nodes in a network increases progressively over time and preferential attachment means the probability of connecting a new node to a node with higher degree is higher than connecting to a node with lower degree. This principle is also known as "rich get richer" principle.

The Python library provides a method $barabasi\_albert$ on a graph to construct a scale free graph as in the following code. This method inputs the number of nodes and the number of edges for a new node. We select 10 nodes and an average of 5 edges per node to build a scale-free graph.

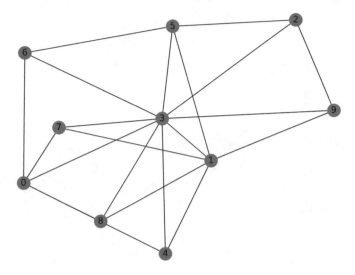

Fig. 10.15 A scale-free network

```
1   import networkx as nx
2   import matplotlib.pyplot as plt
3
4   G=nx.barabasi_albert_graph(10,5)
5   nx.draw(G,with_labels=1)
6   plt.show()
```

A sample output from this code segment shows that nodes 1 and 3 are the hubs with high degrees in the graph of Fig. 10.15

10.8 Kronecker Graphs

The idea of using Kronecker graphs is to construct real-life networks synthetically that have the scale-free and small-world properties [3]. We find that the Kronecker graphs constructed in this manner have power law degree distribution and small diameter as most complex networks. Thus generation of Kroneceker graphs provide us a method to build realistic large complex networks to experiment. Kronecker graphs are built on the matrix operation *Kronecker product*. Building a synthetic network based on this operation recursively provides a graph that obeys real network properties [3].

Kronecker Product of Graphs

Let us first define the Kronecker product of two matrices. Given two matrices $A_{m,n}$ and $B_{m',n'}$, the Kroneker multiplication of A and B is a matrix C of size $(n \cdot n') \times (m \cdot m')$ with elements s below.

$$C = A \otimes B = \begin{bmatrix} a_{11}B & a_{12}B & \dots & a_{1n}B \\ a_{21}B & a_{22}B & \dots & a_{2n}B \\ \vdots & \vdots & \ddots & \vdots \\ a_{m1}B & a_{m2}B & \dots & a_{mn}B \end{bmatrix}$$

Python *numpy* library provides the *kron* method to find the Kronecker product of two matrices. The following Python code shows the product of two matrices using this method.

```
import numpy as np

A = np.array([[1,1,1],[2,2,2],[3,3,3]])
B = np.array([[1,2,3],[4,5,6],[7,8,9]])
C = np.kron(A,B)
print(C)
>>>
[[ 1  2  3  1  2  3  1  2  3]
 [ 4  5  6  4  5  6  4  5  6]
 [ 7  8  9  7  8  9  7  8  9]
 [ 2  4  6  2  4  6  2  4  6]
 [ 8 10 12  8 10 12  8 10 12]
 [14 16 18 14 16 18 14 16 18]
 [ 3  6  9  3  6  9  3  6  9]
 [12 15 18 12 15 18 12 15 18]
 [21 24 27 21 24 27 21 24 27]]
```

Definition 10.5 (*Kronecker product of two graphs*) Let G and H be two graphs with adjacent matrices $A(G)$ and $A(H)$ respectively. The Kronecker product $G \otimes H$ is defined as the graph K with an adjacency matrix formed by the Kronecker product of the adjacency matrices of the two graphs, that is, $A(K) = A(G) \otimes A(H)$.

Based on this definition, a Kronecker graph can be defined as follows.

Definition 10.6 (*Kronecker graph*) A Kronecker graph of order k of a graph G is defined as the kth power of its adjacency matrix $A(G)$ where calculation of power is performed using Kronecker product of matrices. We will call this operation as *Kronecker power*.

The first graph is called the initiator graph. The Python code to build a Kronecker graph with $k = 3$ is shown below where we first find the third Kronecker power of the adjacency matrix of the graph of Fig. 10.16 and then build the graph represented by this adjacency matrix.

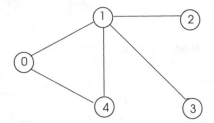

Fig. 10.16 A sample initiator graph

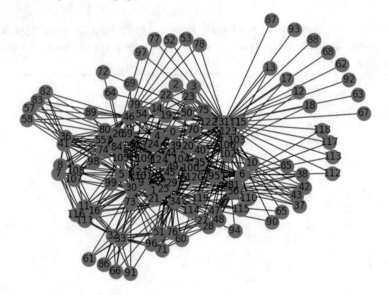

Fig. 10.17 A scale-free network

```
1   import numpy as np
2   import networkx as nx
3   import matplotlib.pyplot as plt
4
5   A = np.array([[0,1,0,0,1],
6                 [1,0,1,1,1],
7                 [0,1,0,0,0],
8                 [0,1,0,0,0],
9                 [1,1,0,0,0]])
10  C = np.kron(A,A)
11  C = np.kron(C,A)
12  G = nx.from_numpy_matrix(C)
13  nx.draw(G,with_labels=1)
14  plt.show()
15  print(C)
```

The plotted graph is depicted in Fig. 10.17 where some complex network properties such as high degree nodes and low diameter are already observable. A Kronecker bipartite graph is built similarly using the adjacency matrix that has vertices of the first vertex set as rows and the vertices of the second as columns.

Kronecker graphs have the power law distribution and small diameter properties found in real networks such as social networks, biological networks and technical networks. We can generate a synthetic Kronecker graph and compare its properties with a real network such as degree distribution and average clustering coefficient.

A *stochastic Kronecker graph* may be formed by creating an $n_1 \times n_1$ probability matrix P_1, computing its kth Kronecker power P_k and for each $p_{uv} \in P_k$, create an edge with probability p_{uv} in the stochastic Kronecker graph [3].

10.9   Chapter Notes

We started this chapter with the main graph models and then reviewed the basic parameters for the analysis of graphs. These parameters provide useful insight to the structure of large graphs consisting of thousands of nodes and tens of thousands of edges or more when visualization of the whole graph is very difficult. We implemented algorithms to evaluate these parameters in Python mainly by using the adjacency or distance matrices of a graph. This way, it is possible to implement these algorithms in a parallel processing environment by simply partitioning the adjacency or the distance matrix to a number of processing elements.

The betweenness values of vertices and edges provide more significant ways of determining the importance of vertices and edges in a network. These measures along with matching index are used to analyze various complex networks such as protein interaction networks and social networks [1].

Exercises and Programming projects

1. Work out the clustering coefficient of the vertices in the graph of Fig. 10.18 by hand and then by implementing the Python algorithm.
2. Find the closeness centralities of the vertices in the graph of Fig. 10.19 by hand and then by implementing the Python algorithm.
3. Find the vertex centralities of the vertices in the graph of Fig. 10.20 by hand and then by implementing the Python method *edge_betweenness*.
4. Write the Python function to find the vertex betweenness centrality of a given graph.

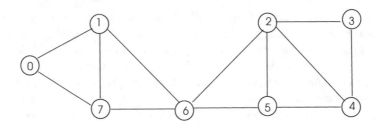

Fig. 10.18 A sample graph for Exercise 1

Fig. 10.19 A sample graph
for Exercise 2

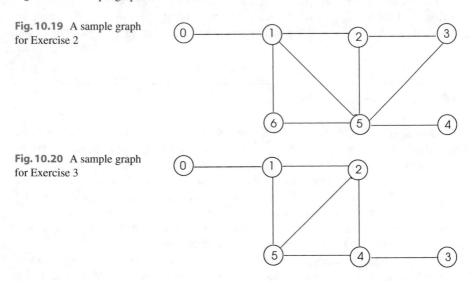

Fig. 10.20 A sample graph
for Exercise 3

References

1. K. Erciyes, *Distributed and Sequential Algorithms for Bioinformatics*. Springer Computational Biology Series (2015)
2. O. Perron, Zur Theorie der Matrices. Mathematische Annalen **64**(2), 248–263 (1907)
3. J. Leskovec, D. Chakrabarti, C. Faloutsos, Z. Ghahramani, Kronecker graphs: an approach to modeling networks. J. Mach. Learn. Res. **11**(2010), 985–1042 (2010)
4. D.J. Watts, S.H. Strogatz, Collective dynamics of small-world networks. Nature **393**, 440–442 (1998)

Graph Partitioning and Clustering

11

Abstract

Partitioning and clustering are two main operations on graphs that find a wide range of applications. Graph partitioning aims at balanced partitions with minimum interactions between partitions. However, graph clustering algorithms attempt to discover densely populated regions of graphs. We review algebraic algorithms for these problems and provide Python implementations of these algorithms in this chapter.

11.1 Graph Partitioning

Graph partitioning problem can be defined as follows: given a graph $G = (V, E, W_V, W_E)$ with,

- V = Set of vertices
- E = Set of edges
- $W_V : V \rightarrow \mathbb{R}$, vertex weights
- $W_E : E \rightarrow \mathbb{R}$, edge weights

graph partitioning is to partition the vertices into k disjoint sets $V_1, \ldots, V_k$ such that,

- $V = \{V_1, \ldots, V_k\}$
- $V_1 \cap V_2 \cap \cdots \cap V_k = \varnothing$

with the aim that the sum of the weights in each partition is approximately equal and the sum of the weights of edges that are in the intersection of vertex sets $V_1, \ldots, V_k$ is minimal. When the vertices have no weights, the partitioning aims at having partitions with almost equal number of vertices and when the graph is unweighted, the goal of the partitioning method is to have a total minimum number of edges between

© Springer Nature Switzerland AG 2021

K. Erciyes, *Algebraic Graph Algorithms*, Undergraduate Topics in Computer Science,

https://doi.org/10.1007/978-3-030-87886-3_11

the partitions. Partitioning a graph has numerous applications; for example, load balancing in a distributed computer system involves sending processes from overloaded nodes to lightly loaded nodes while minimizing communication in various parallel computing applications as we will see.

The problem of choosing an optimal partitioning of a graph is NP-hard [4] and thus, various heuristics are in common use. *Local methods* of graph partitioning assume that the graph is partitioned using some method and refinement of the initial partitioning is then carried out by decreasing the cutsize using another method. *Global methods* typically employ some recursive algorithm to divide the graph into two partitions called *bisection* at each recursion step to arrive at partitions. A global method is commonly used with a local method for improvement.

We will look at three basic graph partitioning methods: the breadth-first-search (BFS) based partitioning and spectral bisection first, then we will review multilevel graph partitioning methods that use graph contraction.

11.1.1   BFS-Based Partitioning

Breadth-first-search (BFS) based partitioning of an unweighted graph is one of the simplest methods to perform this task. The idea is to run the BFS algorithm on a graph from any vertex and record the returned vertices in the order of visit. We then store the first half of visited vertices in one set and the remaining visited vertices in the other. Algorithm 11.1 shows the logic of this algorithm where L is the list of returned vertices from BFS and V_1 and V_2 are the two balanced partitions. The running time of this algorithm is $O(n + m)$ as in the ordinary BFS algorithm as bisection of vertices takes $O(n)$ time.

Algorithm 11.1 *BFS-based Partition*

1: **Input** : $G = (V, E)$ ▷ connected, unweighted graph G
2: **Output** : Partitions V_1 and V_2 of G
3: $V_1 \leftarrow \emptyset, V_2 \leftarrow \emptyset$
4: $L = BFS(v, A)$ ▷ v is any vertex
5: $n = |L|$
6: $k = 0$
7: **while** $k < \lceil n/2 \rceil$ **do**
8:     $V_1 \leftarrow V_1 \cup \{L[i]\}$
9:     $k \leftarrow k + 1$
10: **end while**
11: $V_2 \leftarrow L[k], \ldots, L[n]$

The Python algorithm BFS_Part inputs the adjacency matrix of vertices of a graph and runs the BFS algorithm from our previously coded module BFS of Chapter 7 which returns the neighborhood matrix N, visited nodes and the visited nodes at each level in list $levels$. These nodes at each level are copied to the list *nodes* which is then partitioned into two sets. Note that we could have used the visited node list *vis* which returns the visited nodes in visited order (see Exercise 2).

```
1   ################################################################
2   #               BFS-based Partitioning Algorithm               #
3   ################################################################
4   import numpy as np
5   import BFS as bfs
6
7   def BFS_Part(A,v):
8       n = len(A)
9       V1 = []
10      V2 = []
11      nodes = []            # list to hold nodes
12      N, vis, levels = bfs.BFS(A,v)  # run BFS
13      for i in range(0,len(levels)):# store levels
14          nodes.extend(levels[i])    # in list
15      V1 = np.copy(nodes[0:int(n/2)]) # divide nodes
16      V2 = np.copy(nodes[int(n/2):n]) # by 2
17      return V1, V2
18
19  if __name__ == '__main__':
20      B = np.array([[1,1,0,0,0,0,1,0,0,0], [1,1,1,0,0,0,1,0,0,0],
21          [0,1,1,1,1,1,1,0,0,0], [0,0,1,1,0,0,0,0,1,1],
22          [0,0,1,0,1,1,0,0,0,1], [0,0,1,0,1,1,1,0,0,0],
23          [1,1,1,0,0,1,1,1,0,0], [0,0,0,0,0,0,1,1,0,0],
24          [0,0,0,1,0,0,0,0,1,0], [0,0,0,1,1,0,0,0,0,1]],
25          dtype=bool)
26      V1, V2 = BFS_Part(B,0)
27      print ("Partitions: V1:", V1, "V2:",V2)
28  >>>
29  Partitions: V1: [0 1 6 2 5] V2: [7 3 4 8 9]
```

Running this algorithm in the graph of Fig. 11.1 resulted in the BFS tree shown in bold arrows and the partitions shown.

11.1.2 Spectral Bisection

Spectral bisection is a method of partitioning a graph into two sets of vertices based on its algebraic properties. The Laplacian matrix L of a graph was defined as $L = D - A$ where D is the degree matrix having the degrees of the vertices along its diagonal and A is the adjacency matrix of the graph. This matrix consists of elements l_{ij} such that,

$$l_{ij} = \begin{cases} -1 & \text{if } i \neq j \text{ and } a_{ij} = 1 \\ 0 & \text{if } i \neq j \text{ and } a_{ij} = 0 \\ -d_{ij} & \text{if } i = j \end{cases}$$

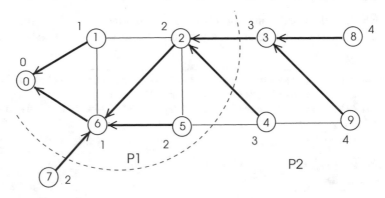

Fig. 11.1 A sample graph

where d_{ii} is the diagonal element of matrix D. The matrix L has some interesting properties. The multiplicity of the eigenvalues that are equal to zero in L gives the number of components of the graph as noted. Moreover, the eigenvector associated with the second smallest eigenvalue of L called *Fiedler vector* can be used to partition a graph [3]. The nodes corresponding to negative values in this vector may be included in one partition and nodes that are associated with positive values in the other. An algorithm to partition a graph using Fiedler vector property is shown in Algorithm 11.2 where Fiedler vector is first computed and the partitions are then formed accordingly.

Algorithm 11.2 *Spectral Bisection*

1: **Input** : Adjacency matrix A of a graph $G = (V, E)$ ▷ connected, unweighted graph G
2: **Output** : Partitions V_1 and V_2 of G
3: $V_1 \leftarrow \emptyset, V_2 \leftarrow \emptyset$
4: **compute** degree matrix D
5: $L = D - A$
6: **compute** eigenvalues and eigenvectors of L
7: let e_2 be the second smallest eigenvalue of L
8: let c_2 be the eigenvector corresponding to e_2
9: **for all** $x \in c_2$ **do**
10:     **if** $x > 0$ **then**
11:         $V_1 \leftarrow V_1 \cup \{x\}$
12:     **else if** $x > 0$ **then**
13:         $V_2 \leftarrow V_2 \cup \{x\}$
14:     **else**
15:     **end if**
16: **end for**

The Python program to implement this algorithm inputs the adjacency matrix A and forms the diagonal matrix D by summing each row of matrix A in lines 14–15.

The Laplacian matrix L is then computed and its eigenvalues are stored in arrays *eigvals* and *eigvecs* using the *eig* function from the library *scipy* in lines 16–20. The second smallest value of *eigvals* is then found by sorting eigenvalues and the elements of the eigenvector with this value are tested; negative values are stored in partition V_1 and others in V_2.

```python
##############################################################
#                Spectral Bisection Algorithm               #
##############################################################

import numpy as np
import scipy.linalg as la
import statistics as st

def Spectral_Bisection(A):
    n=len(A)
    V1 = []      # first partition
    V2 = []      # second partititon
    D = np.zeros((n,n)) # build degree matrix
    for i in range(0,n):
        D[i,i] = sum(A[i,:])
    L = D - A   # compute L
    eigvals, eigvecs = la.eig(L) # find eigens of L
    eigvals = eigvals.real
    eigvecs = eigvecs.real
    eigens = eigvals.copy()
    eigvals.sort()                      # find second smallest
    ind_2 = np.argwhere(eigens==eigvals[1])
    ind_2 = int(ind_2)
    for i in range(0,n):
        eigvecs[ind_2][i] = round(eigvecs[8][i],3)
    for i in range(0,n):                # partition
        if eigvecs[ind_2,i] < 0:
            V1.append(i)
        elif eigvecs[ind_2,i] > 0:
            V2.append(i)
        else:
            if len(V1) < len(V2):
                V1.append(i)
            else:
                V2.append(i)
    return V1, V2

if __name__ == '__main__':

    B = np.array([[0,1,0,0,0,0,0,0,0,1,0,0,0,0,0],
                  [1,0,1,0,0,0,0,0,0,1,1,1,0,0,0],
                  [0,1,0,1,0,0,0,0,1,1,0,0,0,0,0],
                  [0,0,1,0,1,0,1,1,0,0,0,0,0,0,0],
                  [0,0,0,1,0,1,1,0,0,0,0,0,1,0,0],
                  [0,0,0,0,1,0,0,0,0,0,0,0,0,0,0],
```

```
46              [0,0,0,1,1,0,0,1,0,0,0,0,0,0,0],
47              [0,0,0,1,0,0,1,0,0,0,0,0,0,0,0],
48              [0,0,1,0,0,0,0,0,0,1,0,0,0,0,0],
49              [1,1,1,0,0,0,0,0,1,0,0,0,0,1,1],
50              [0,1,0,0,0,0,0,0,0,0,0,0,0,0,0],
51              [0,1,0,0,0,0,0,0,0,0,0,0,0,0,0],
52              [0,0,0,0,1,0,0,0,0,0,0,0,0,0,0],
53              [0,0,0,0,0,0,0,0,0,1,0,0,0,0,0],
54              [0,0,0,0,0,0,0,0,0,1,0,0,0,0,0]])

56      V1, V2 = Spectral_Bisection(B)
57      print ("Partitions: V1", V1,"V2",V2)
58   >>>
59   Partitions V1: [3, 5, 9, 10, 12, 13, 14]
60              V2: [0, 1, 2, 4, 6, 7, 8, 11]
```

Running of this algorithm on the sample graph of Fig. 11.2 resulted in the partitions shown with partition V_1 enclosed in the curve and partition V_2 vertices with double circles. This algorithm may be used recursively to provide partitions greater than 2 and as a subroutine for more sophisticated partitioning algorithms.

11.1.3   Multilevel Graph Partitioning

A multilevel graph partitioning method builds smaller graphs from the initial graph by coarsening recursively, and when the small graph is small enough, partitions the small graph is partitioned and uncoarsened to the original graph. This method mainly consists of coarsening, partitioning and uncoarsening steps as below [5]:

1. *Coarsening*: Obtain a sequence of smaller graphs $G_i = (V_i, E_i)$, $i = 1, \ldots, k$ from the original graph $G_0 = (V_0, E_0)$.

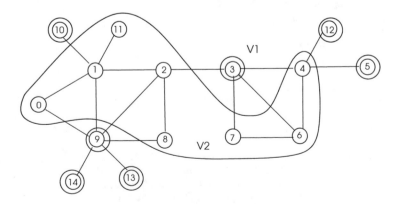

Fig. 11.2 A sample graph to test spectral bisection algorithm

2. *Partitioning*: Partition the smallest graph G_k using a partitioning algorithm such as recursive spectral bisection.
3. *Uncoarsening*: Grow the smaller graphs back to original by keeping the partitions and using refinement such as Kernighan–Lin algorithm at each step.

The partitioning step may be performed using a heuristic such as random matching (RM) with the following steps.

1. **Input**: $G = (V, E)$
2. **Output**: Maximal matching M
3. $M \leftarrow \varnothing$
4. **while** $E \neq \varnothing$
5.   Select a random unmatched vertex u
6.   Select an edge (u, v) incident to u randomly
7. $M \leftarrow M \cup \{(u, v)\}$
8.   Remove all edges incident to u and v from E

This algorithm is similar to a random matching algorithm where edges are selected at random, but here, we select an unmatched vertex at random rather than an edge. A different heuristic called heaviest edge matching (HEM) is similar to RM but the heaviest weight edge is selected instead of a random edge in step 5 of the above algorithm. A variation of HEM sorts the edges from highest to smallest with respect to their weights and then selects edges from this list in sequence without violating matching property.

Partitioning of a sample graph using sorted HEM is depicted in Fig. 11.3, we stop when the number of partitions is 4 in (a) and project the partitioned graph to its original form in (c). The partitions formed are P_{12} with 4 nodes, P_{34} with 3 nodes, P_5 with 2 nodes and P_{67} with 3 nodes and a total weight of inter-partition edges.

11.2 Graph Clustering

Clustering is the process of grouping of closely related data elements in order to identify the structure in data. The groups that contain closely related elements are called *clusters*. *Graph clustering* is the process of grouping the vertices that are in vicinity of each other such that the density of edges in a cluster is significantly higher than the density between the clusters.

A direct method to access the quality of a cluster obtained after a graph clustering method is to compare the density of edges inside a cluster to the average graph density. Edges inside a cluster are called the *internal edges* and the edges between the clusters are called *external edges*. The *intracluster density* of a cluster C_i is defined as the ratio of the number of internal edges in C_i to the maximum possible number of edges in C_i. A graph with three clusters is depicted in Fig. 11.4.

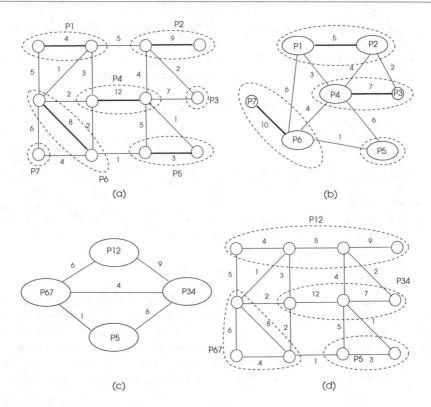

Fig. 11.3 Multilevel graph partitioning using HWEM

11.2.1 MST-Based Clustering

A minimum spanning tree of a weighted graph may be used for clustering with the
following logic: a tree is acyclic and every edge of a tree is a bridge. Thus, removing
an edge from the MST of a graph G divides G into two clusters and continuing in
this manner provides clusters. We need to remove $k - 1$ heaviest edges from the
MST of the graph to obtain k clusters. Since it is difficult to guess the value of k,
a threshold τ may be used such that all edges that have a weight larger than τ are
removed from the MST to form the clusters. The following steps of this algorithm
can now be stated.

1. **Input**: A weighted graph $G = (V, E, w)$, threshold τ.
2. **Output**: $\mathcal{C} = \{C_1, C_2, \ldots, C_k\}$, k clusters of G.
3. Find MST T of G.
4. Remove all edges from G that have a weight larger than τ.
5. Find the components of G_N which are the clusters of G.

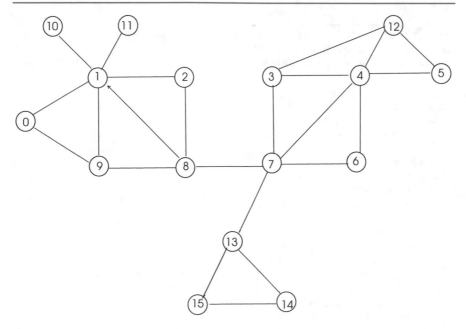

Fig. 11.4 Clusters of a sample graph

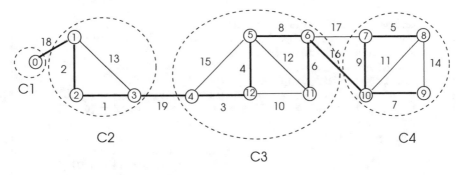

Fig. 11.5 MST-based clustering of a sample graph

Python Implementation

The Python code shown below first constructs the MST T of the given weighted graph of Fig. 11.5 and then iteratively removes the heaviest edge from T until the given number of 4 clusters is obtained. Note that converting this algorithm to test edge weights of T that have larger value than τ can be simply performed (See Exercise 2).

```
1   ################################################################
2   #                MST-based Clustering Algorithm                #
3   ################################################################
4
5   import numpy as np
6   import math
7   import JPRIM as mstr
8   import BFS as bfs
9
10  def MST_Cluster(D):
11    n=len(D)
12    T = np.copy(D)
13    MST_graph = np.zeros((n,n))
14    C = []
15    n_cluster = 4
16    d, mst = mstr.Prim(T,0)      # find MST
17    for i in range(0,len(mst)):
18        MST_graph[i,mst[i]] = D[i,mst[i]] # form MST graph
19        MST_graph[mst[i],i] = D[mst[i],i]
20        MST_graph[i,i] = 1
21    MST_graph[0,0] = 1
22    k = 1
23    while k < n_cluster:      # iterate until n_clusters
24        CT = []
25        max_val = np.amax(MST_graph)
26        max_ind = np.where(MST_graph==np.amax(MST_graph))
27        MST_graph[max_ind[0][0],max_ind[0][1]]=0
28        MST_graph[max_ind[1][0],max_ind[1][1]]=0
29        MST_logic = MST_graph.astype(bool) # run BFSs
30        N, verts1, levels1 = bfs.BFS(MST_logic, max_ind[0][0])
31        N, verts2, levels2 = bfs.BFS(MST_logic, max_ind[0][1])
32        CT.append(verts1)
33        CT.append(verts2)
34        for i in range(0,len(C)):  # update cluster list
35            if max_ind[0][0] in C[i] or max_ind[0][1] in C[i]:
36                del C[i]
37                break
38        C.extend(CT)              # add visited vertices to clusters
39        k = k+1
40    return  C
41
42  if __name__ == '__main__':
43    x = float('inf')
44    B = np.array([[x,18,x,x,x,x,x,x,x,x,x,x,x],
45                  [18,x,2,13,x,x,x,x,x,x,x,x,x],
46                  [x,2,x,1,x,x,x,x,x,x,x,x,x],
47                  [x,13,1,x,19,x,x,x,x,x,x,x,x],
48                  [x,x,x,19,x,15,x,x,x,x,x,x,3],
49                  [x,x,x,x,15,x,8,x,x,x,x,12,4],
50                  [x,x,x,x,x,8,x,17,x,x,16,6,x],
```

```
51              [x,x,x,x,x,x,17,x,5,x,9,x,x],
52              [x,x,x,x,x,x,x,5,x,14,11,x,x],
53              [x,x,x,x,x,x,x,x,14,x,7,x,x],
54              [x,x,x,x,x,x,16,9,11,7,x,x,x],
55              [x,x,x,x,x,12,6,x,x,x,x,x,10],
56              [x,x,x,x,3,4,x,x,x,x,x,10,x]])
57
58    clusters = MST_Cluster(B)
59    print("Clusters:", clusters)
60  >>>
61  Clusters:  [[0],  [1,  2,  3],  [6,  5,  11,  12,  4],  [10,  7,  9,  8]]
```

An MST of a sample graph is depicted in Fig. 11.5 in bold lines. Since the edge weights are distinct, this MST is unique. Running of the Python algorithm on this graph yields the clusters shown in dashed circles in the figure as expected since the edges joining these clusters are the heaviest weight edges in the MST. The removal of edges will be in the order of edges having weights 19, 18 and 16 to result in these four clusters.

11.2.2 Shared Nearest Neighbor Clustering

The number of shared neighbors between a vertex pair (u, v) can be used to form clusters by placing the vertices with significant number of common neighbors in the same cluster. This method is known as *shared nearest neighbor* (SNN) clustering which is based on the concept that if two vertices have many common neighbors, they should be in the same cluster. The steps of this algorithm may be formed as follows [5].

1. **Input**: An unweighted graph $G = (V, E)$.
2. **Output**: $C = \{C_1, C_2, \ldots, C_k\}$, k clusters of G.
3. Find the number of common neighbors for each vertex pair $(u, v) \in E$.
4. Transform G into the weighted graph G_N where each edge (u, v) denotes the number of common neighbors of vertices u and v.
5. Remove all edges from G_N that have a weight less than a threshold τ.
6. Find the components of G_N which are the clusters of G.

If the graph is weighted, it can be transformed into an unweighted one by first removing any edge that has a weight less than a threshold from the graph.

Python Implementation
We will transform the above algorithm steps to the following Python code. The number of common neighbors for each vertex pair is first determined and the adjacency matrix of the weighted graph GN is formed in lines 16–25. Applying the steps of this algorithm in the graph of Fig. 11.6 results in the clusters shown. We apply component finding algorithm of Chap. 8 to find clusters.

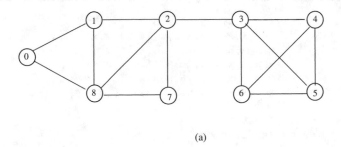

(a)

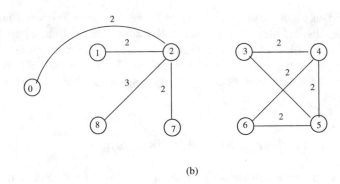

(b)

Fig. 11.6 A sample graph to test SNN algorithm

```
1   ################################################################
2   #              Shared Nearest Neighbor Algorithm               #
3   ################################################################
4
5   import numpy as np
6   import COMP as comp
7
8   def Shared_NN(A,u,v):
9       n = len(A)
10      N_u = []
11      N_v = []
12      GN = np.zeros((n,n))
13      n=len(A)
14      n_common = 0
15
16      for j in range(0,n):        # find neighbors of u
17              if A[u,j]==1:
18                  N_u.append(j)
19          if v in N_u:
20              N_u.remove(v)
21      for j in range(0,n):        # find neighbors of v
22              if A[v,j]==1:
23                  N_v.append(j)
```

```
24       if u in N_v:
25          N_v.remove(u)
26
27       for i in range(0,len(N_u)): # find common neighbors
28          for j in range(0,len(N_v)):
29             if N_u[i] == N_v[j]:
30                n_common = n_common + 1
31       return n_common
32
33    if __name__ == '__main__':
34
35       B = np.array([[0,1,0,0,0,0,0,0,1],
36                     [1,0,1,0,0,0,0,1,1],
37                     [0,1,0,1,0,0,0,1,1],
38                     [0,0,1,0,1,1,1,0,0],
39                     [0,0,0,1,0,1,1,0,0],
40                     [0,0,0,1,1,0,1,0,0],
41                     [0,0,0,1,1,1,0,0,0],
42                     [0,0,1,0,0,0,0,0,1],
43                     [1,1,1,0,0,0,0,1,0]])
44       n = len(B)
45       GN = np.zeros((n,n))    # initialize GN
46       for i in range (0,n):
47          for j in range (i+1,n):
48             GN[i,j] = Shared_NN(B,i,j) # form GN
49             GN[j,i] = GN[i,j]
50       print("GN:")
51       print(GN)
52       for i in range (0,n):
53          for j in range (0,n): # remove light edges
54             if GN[i,j] < 2:
55                GN[i,j] = 0
56             if i == j:
57                GN[i,i] = 1
58       print("GN with removed edges:")
59       print(GN)
60       print("")
61       clusters = comp.Component_Find(GN.astype(bool)) # find clusters
62       print ("Clusters:",clusters)
63    >>>
64    GN:
65    [[0. 1. 2. 0. 0. 0. 0. 1. 1.]
66     [1. 0. 2. 1. 0. 0. 0. 2. 3.]
67     [2. 2. 0. 0. 1. 1. 1. 1. 2.]
68     [0. 1. 0. 0. 2. 2. 2. 1. 1.]
69     [0. 0. 1. 2. 0. 2. 2. 0. 0.]
70     [0. 0. 1. 2. 2. 0. 2. 0. 0.]
71     [0. 0. 1. 2. 2. 2. 0. 0. 0.]
72     [1. 2. 1. 1. 0. 0. 0. 0. 1.]
73     [1. 3. 2. 1. 0. 0. 0. 1. 0.]]
74    GN with removed edges:
75    [[1. 0. 2. 0. 0. 0. 0. 0. 0.]
76     [0. 1. 2. 0. 0. 0. 0. 2. 3.]
77     [2. 2. 1. 0. 0. 0. 0. 0. 2.]
78     [0. 0. 0. 1. 2. 2. 2. 0. 0.]
79     [0. 0. 0. 2. 1. 2. 2. 0. 0.]
```

```
80    [0. 0. 0. 2. 2. 1. 2. 0. 0.]
81    [0. 0. 0. 2. 2. 2. 1. 0. 0.]
82    [0. 2. 0. 0. 0. 0. 0. 1. 0.]
83    [0. 3. 2. 0. 0. 0. 0. 0. 1.]]
84
85   Clusters: {0: [0, 2, 1, 8, 7], 1: [3, 4, 5, 6]}
```

11.2.3 Edge Betweenness Clustering

The edge betweenness value of an edge (u, v) in a graph was defined as the ratio of the number of shortest paths that pass through edge (u, v) to total number of shortest paths as obtained by an all-pairs-shortest-paths algorithm such as Floyd–Warshall. If this value is comparatively higher than other edges in a given graph, the probability of such an edge joining clusters of a graph is high. Based on this observation, edge betweenness clustering method eliminates the edges with a betweenness value higher than a given threshold τ from the graph and the components of the resulting graph may be discovered to yield the clusters as given by the following steps of the algorithm.

1. **Input**: An unweighted graph $G = (V, E)$.
2. **Output**: $C = \{C_1, C_2, \ldots, C_k\}$, k clusters of G.
3. Calculate edge betweenness value $\forall (u, v) \in E$.
4. Remove all edges from G that have a higher betweenness value than a threshold τ.
5. Find the components of G_N which are the clusters of G.

Python Implementation
We will implement the above steps of the algorithm in Python using the *networkx* method *edge_betweenness* to calculate betweenness values for edges.

```
1    import numpy as np
2    import networkx as nx
3    import matplotlib.pyplot as plt
4    import COMP as comp
5
6    A = np.array([   [0,1,0,0,0,0,0,0,1],
7                     [1,0,1,0,0,0,0,1,1],
8                     [0,1,0,1,0,0,0,1,1],
9                     [0,0,1,0,1,1,1,0,0],
10                    [0,0,0,1,0,1,1,0,0],
11                    [0,0,0,1,1,0,1,0,0],
12                    [0,0,0,1,1,1,0,0,0],
13                    [0,0,1,0,0,0,0,0,1],
14                    [1,1,1,0,0,0,0,1,0]])
15   G = nx.Graph(A)
16   nx.draw(G, with_labels=1)
17   plt.show()
18
```

```
19  threshold = 0.4
20  E = nx.edge_betweenness_centrality(G)
21  for key,value in E.items():
22      E[key] = round(E[key],2)
23      if value > threshold:
24          A[key[0],key[1]] = 0
25          A[key[1],key[0]] = 0
26  print(E)
27  G = nx.Graph(A)
28  nx.draw(G, with_labels=1)
29  plt.show()
30  n = len(A)
31  for i in range (0,n):
32      A[i,i] = 1
33  print("")
34  clusters = comp.Component_Find(A.astype(bool)) # find clusters
35  print ("Clusters:",clusters)
36  >>>
37  {(0, 1): 0.11, (0, 8): 0.11, (1, 2): 0.21, (1, 7): 0.04,
38   (1, 8): 0.03, (2, 3): 0.56, (2, 7): 0.14, (2, 8): 0.21,
39   (3, 4): 0.17, (3, 5): 0.17, (3, 6): 0.17, (4, 5): 0.03,
40   (4, 6): 0.03, (5, 6): 0.03, (7, 8): 0.04}
41
42  Clusters: {0: [0, 1, 8, 2, 7], 1: [3, 4, 5, 6]}
```

The running of this algorithm on a simple graph of Fig. 11.7 results in two clusters as output by the program along with edge betweenness values of edges.

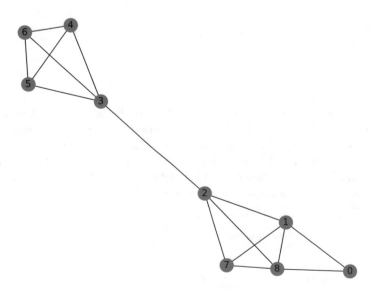

Fig. 11.7 A sample graph to test edge-betweenness algorithm

11.2.4 k-Core Clustering

An V' induced subgraph $G' = (V', E')$ of a graph $G = (V, E)$ is a k-core of graph G if and only if every vertex v of V' has a degree greater than or equal to k and G' is the maximum graph that has this property [2]. The main core of a graph G is a core of G that has the highest number of vertices. The highest core number is denoted by the *core value* of a vertex.

We can detect dense regions of a graph by discovering its cores since a k-core will have vertices with at least k degree. Finding k-cores can be accomplished by Batagelj ad Zaversnik algorithm [1] shown in Algorithm 11.3 [2]. Vertices are sorted with respect to their degrees in increasing order to a queue Q first. Then, vertex v is dequeued from Q is labeled with core value of its degree and degrees of all of its neighbors are decremented if they have a degree larger than the degree (MORE EXPLAIN). The running time of this algorithm is $O(max(m, n))$ [1].

Algorithm 11.3 *Batagelj Algorithm*

1: **Input** : $G = (V, E)$ ▷ a connected graph G
2: **Output** : $cores[n]$ ▷ core values of vertices
3: Sort vertices with respect to increasing degrees into Q
4: **while** $Q \neq \emptyset$ **do**
5:     $u \leftarrow deque(Q)$
6:     $cores[v] \leftarrow deg(v)$
7:     **for all** $u \in N(v)$ **do**   ▷ update neighbor degrees
8:         **if** $deg[u] > deg(v)$ **then**
9:             $deg(u) \leftarrow deg[u] - 1$
10:        **end if**
11:    **end for**
12:    update vertex degrees in Q
13: **end while**

Python Implementation

A straightforward implementation of this algorithm in Python is shown below where we first calculate degrees of a given graph G and sort the vertices with degrees into the list *degs_sorted* and then, we pop out a vertex from the front of this list, assign its core value as its degree and decrement all neighbor degrees as in Algorithm 11.3. We need to consider starting a new core by testing whether the degree of the dequeued vertex is different than the current core value in lines 22–26.

```
1   ##########################################################################
2   #                        k-core Algorithm                               #
3   ##########################################################################
4
5   import numpy as np
6
7   def Find_Cores(A):
8       n=len(A)
9       cores = {}
10      core_temp = []
11      core_vals = [-1]*n
```

```
12      D = np.sum(A,axis=1)            # calculate degrees
13      degs = []
14      for i in range(0,n):
15          degs.append([i,D[i]])       # form list
16      degs_sorted = sorted(degs, key=lambda x:x[1]) # sort list
17      k = 0
18      last = degs_sorted[0][1]
19      length = len(degs_sorted)
20      while length > 0:
21          v,deg = degs_sorted.pop(0) # dequeue first vertex
22          core_vals[v] = deg          # assign its core
23          if deg != last:             # start new core
24              last = deg
25              core_temp = []
26              k = k+1
27          core_temp.append(v)         # append vertex to core list
28          cores[k] = core_temp        # update cores
29          for i in range(0,n):        # update degrees
30              if A[v,i] == 1:
31                  for j in range(0,len(degs_sorted)):
32                      if degs_sorted[j][0] == i and degs_sorted[j][1]>deg:
33                          degs_sorted[j][1] = degs_sorted[j][1]-1
34          degs_sorted = sorted(degs_sorted, key=lambda x:x[1])
35          length = length -1
36      return  cores
37
38  if __name__ == '__main__':
39      B = np.array([[0,1,0,0,0,0,0,0,0,0,0,0],
40                    [1,0,1,1,0,0,0,0,0,0,0,0],
41                    [0,1,0,1,0,0,0,0,0,0,0,0],
42                    [0,1,1,0,1,0,0,0,0,0,0,0],
43                    [0,0,0,1,0,1,0,0,0,1,1,1],
44                    [0,0,0,0,1,0,1,0,0,1,1,0],
45                    [0,0,0,0,0,1,0,1,0,1,1,0],
46                    [0,0,0,0,0,0,1,0,0,0,0,0],
47                    [0,0,0,0,0,0,0,0,0,1,0,0],
48                    [0,0,0,0,1,1,1,0,1,0,1,0],
49                    [0,0,0,0,1,1,1,0,0,1,0,0],
50                    [0,0,0,1,0,0,0,0,0,0,0,0]])
51      core_vals = Find_Cores(B)
52      print("Cores:", core_vals)
53  >>>
54  Cores: {0: [0, 7, 8, 11], 1: [2, 1, 3], 2: [6, 4, 5, 10, 9]}
```

Updating degrees of neighbors is the most time consuming part of this algorithm with two nested *for* loops in lines 28–35. However, this part of the code can be performed in parallel by distributing the queue elements to p processors which update their queue portions and a master process may then merge all these portions. Running of this algorithm in the sample graph of Fig. 11.8 produced the cores shown.

11.3   Chapter Notes

We reviewed two important methods for the analysis of large graphs: graph partitioning and graph clustering. Graph partitioning basically is used to divide a graph into balanced subgraphs. These subgraphs have similar number of vertices and minimal number of edges between them in unweighted graph partitioning. On the other hand, we search for approximately equal total weights of vertices and a total mini-

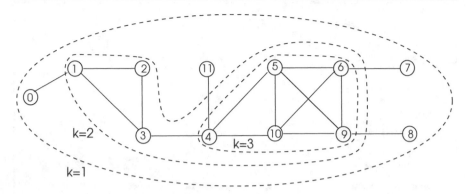

Fig. 11.8 k-cores of a sample graph

mum weight of edges in a graph that has weighted vertices and edges. In both cases, the graph partitioning problem is NP-hard and various heuristics are used. A BFS-based partitioning for an unweighted graph is one of the basic partitioning methods and spectral bisection may serve as the basis for more sophisticated partitioning algorithms. Multi-layer partitioning is a commonly used method in practice for this problem. A graph is coarsened using some heuristic into smaller and smaller graphs and the smallest graph is then partitioned with much ease than partitioning the original graph. The partitioned graph is then projected back to the original graph with some refinement along the process. Commonly used heuristics in this method are matching and spectral bisection.

Graph clustering has a different meaning than partitioning, in which detection of existing clusters in a graph is sought. These clusters will not be balanced in general and it is difficult to estimate their size before the clustering algorithm. We reviewed some of the main clustering algorithms which are MST-based clustering, edge betweenness-based clustering and k-core algorithm which are used in large biological networks [2].

There are numerous studies and published results in both graph partitioning and clustering as there is not a single method that suits all applications in this ongoing area of research.

Exercises and Programming Projects

1. Modify the BFS-based graph partitioning algorithm in Python such that the returned list of visited nodes from the BFS algorithm is divided into two partitions. Run this algorithm in the graph of Fig. 11.9 to obtain two partitions.
2. Modify the spectral graph partitioning algorithm in Python such that we can have k partitions instead of 2. Run this algorithm in the graph of Fig. 11.10 to obtain four partitions.

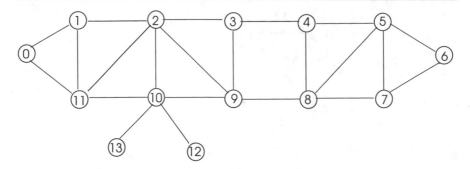

Fig. 11.9 A sample graph for Exercise 1

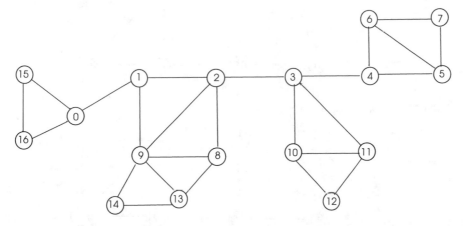

Fig. 11.10 A sample graph for Exercise 2

3. Modify the MST-based clustering algorithm Python code such that clusters are formed using a threshold value τ. Run this algorithm in the graph of Fig. 11.11 with a τ value of 20.
4. Find the k-cores of the graph shown in Fig. 11.12 by running the Python k-cores algorithm and confirm by manual calculation.

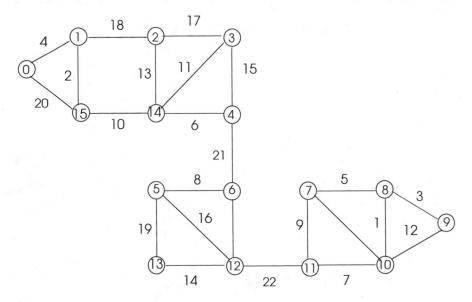

Fig. 11.11 A sample graph for Exercise 3

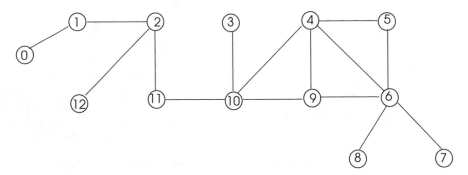

Fig. 11.12 A sample graph for Exercise 4

References

1. V. Batagelj, M. Zaversnik, An O(m) algorithm for cores decomposition of networks. CoRR (Computing research repository) (2003). arXiv:0310049
2. K. Erciyes, *Guide to Graph Algorithms: Sequential, Parallel and Distributed.* Springer Texts in Computer Science (2018)
3. M. Fiedler, Algebraic connectivity of graphs. Czech. Math. J. **23**, 298–305 (1973)
4. M. Garey, D. Johnson, L. Stockmeyer, Some simplified NP-complete graph problems. Theor. Comput. Sci. **1**, 237–267 (1976)
5. R. A. Jaruis, E. A. Patrick, Clustering using a similarity measure based on shared nearest neighbors. IEEE Trans. Comput. **22**(11), 1025–1034 (1973)

Index

© Springer Nature Switzerland AG 2021
K. Erciyes, *Algebraic Graph Algorithms*, Undergraduate Topics in Computer Science,
https://doi.org/10.1007/978-3-030-87886-3

Printed in the United States
by Baker & Taylor Publisher Services